H. Grönvold pinxit. Pawson & Brailsford Lith. Sheffield.

PASSERIFORMES: HIRUNDINIDÆ, MUSCICAPIDÆ.

The Swallow

A.Thorburn

THE SWALLOW

A Biography

A year in the life of
the world's best-known,
and best-loved, bird

Stephen Moss

SQUARE PEG

1 3 5 7 9 10 8 6 4 2

Square Peg, an imprint of Vintage,
20 Vauxhall Bridge Road,
London SW1V 2SA

Square Peg is part of the Penguin Random House group of companies whose addresses can be found at global.penguinrandomhouse.com

First published by Square Peg in 2020

Penguin.co.uk/vintage

A CIP catalogue record for this book is available from the British Library

ISBN 9781529110265

Typeset by Dinah Drazin

Printed and bound in China by C&C Offset Printing Co., Ltd.

To Shelley, Luke, Kaya and Skie – our family's very own migrants to the southern hemisphere

Fröhliche
Pfingsten

My little book is concerned with the life history of a bird which is beyond doubt the best known, and certainly the best loved, species in the world – the swallow.

Collingwood Ingram,
The Migration of the Swallow (1974)

AUTHOR'S NOTE

The 'swallow' featured in this book is *Hirundo rustica*, officially known as the 'barn swallow' to avoid confusion with the eighty or so other species in its family. However, for convenience I mostly refer to it as 'the swallow', except where that might cause confusion.

Contents

PROLOGUE

It's a year almost that I have not seen her:
Oh, last summer green things were greener,
Brambles fewer, the blue sky bluer.

It's surely summer, for there's a swallow …

Christina Rossetti, 'A Bird Song' (1862)

PROLOGUE

The Traveller Returns

It's early April, and the village is waiting.

The paperboy, yawning as he tries to stay awake; the farmer, herding his sheep from one field to another; the woman, walking her dog along the lane – all are waiting – though they may not yet realise what they are waiting for. I am waiting too, in this little Somerset village, where a few wheezy sparrows greet the cold, grey dawn.

All of us are expecting a visitor. One which, more than any other, heralds the end of winter and the coming of spring. Not a human traveller, but a bird: the first swallow.

We are not the only ones doing so. All across the Northern Hemisphere, from Alaska to New York, Ireland to Japan, people are gazing up at the sky for the first glimpse of their annual guest.

Our swallow, along with tens of millions of others, is already on his way back. Back from his African winter quarters, after one of the longest journeys made by any living creature on earth. Back home.

By the time he arrives, he will have flown more than 6,000 miles, across land and sea, to reach the place where he hatched out of an egg not so very long ago. How he does so is one of the world's greatest natural wonders; little short of a miracle.

This swallow's journey begins two months earlier, as he swoops and dives through the humid air, hawking for insects over the dry South African savannah. It's the height of the southern summer, and for this swallow, and millions of his companions, it appears to be the ideal place to live: with endless sunshine, plenty of insect food, and water to drink and bathe in.

Yet our swallow is beginning to feel restless. His instincts are telling him to leave this wonderland; to embark on a journey that will test his skill, stamina and endurance; a journey he may never complete. He has a good chance of dying before achieving his goal.

But why leave at all? To answer this question, we must go back to when this swallow's distant ancestors left Africa and headed north for the very first time. What impelled them we do not know. Perhaps the skies there were simply too crowded, and they were seeking out new places to feed and breed. Perhaps a genetic mutation created an instinctive wanderlust, an urge to explore new horizons.

What we do know is that, once the travelling habit took hold, these swallows could never stay put again. Generation after generation, over tens of thousands of years, they pushed further and further north, until the descendants of those early pioneers finally made it all the way to the land we now call Europe.

Here, as the last of the Ice Ages finally retreated, they found another ideal place to live: longer hours of daylight than in

Africa, a seasonal abundance of flying insects on which to feed, and safe and plentiful nesting sites. They could raise a family without having to compete with the many other swallow species back in their ancestral home.

But once summer was over, and the winds of autumn began to blow, that abundance of insects soon dwindled to nothing. Faced with the inevitability of starvation and death, these birds then embarked on the long journey back across the Equator to southern Africa, where they would spend the remaining half of the year.

So, although we like to think of these birds as 'our' swallows, the truth is rather different. These birds belong to Africa. They visit us for a few short months each spring and summer: a brief but welcome stay. In return, we provide them with food and lodging, cherish their presence while they are with us, and miss them when they are gone.

By the end of February, our swallow can no longer ignore the urge to migrate. Tomorrow, at dawn, he will set off on the first stage of his long journey. But now he has a more urgent need: a safe place to spend the night.

The sun sinks fast at these latitudes, and the grassy plains are soon bathed in the last, rapidly dying rays of daylight. By Kipling's great, grey-green, greasy Limpopo River there stands ... not an Elephant's Child, but a vast fig tree, its silver branches etched against the deep red sky. In its shadow stretches a field of maize, gold and glowing, where a flock of swallows has come to roost for the night. Ours is among them. One by one, the birds drop like spots of rain from the darkening sky into the tall crops, seeking safety in numbers. From a sea of gently waving grass, nightjars and crickets sing a rough-edged evensong; a prelude to the night's activity. But for our swallow, darkness brings no quiet rest.

Swallows never really experience uninterrupted sleep. Instead, their heart rate slows down, and they snatch brief catnaps, always alert to the possibility of danger. Tonight, hyenas whoop to one another across the dry plains. They've heard the lions roar, and they know there might be a free meal on offer, somewhere out there in the darkness.

As the cold blue light signals another African dawn, a pair of fish eagles greet the day from their high perch above the water. These birds are the king and queen of this land, magnificent creatures whose snow-white head-dress gives them a regal air of authority. They stare down at the river with piercing dark eyes. A trail of bubbles is breaking on the surface.

These are produced by a family of hippos, barely visible beneath the surface. Perhaps they are responding to the eagles' cries. Or maybe, as local people believe, the hippos are telling each other jokes and, when they can restrain their laughter no longer, it erupts in a stream of bubbles.

The swallow stretches his wings, shivers briefly, and sets off, his long wings scything through the chilly air. It's time to leave.

By the time the land is warming up he has already flown many miles north, with only a few tiny morsels, grabbed in mid-air, to satisfy his appetite. Now he needs to refuel with something more substantial. Breakfast.

In the far distance, a mile or so away, are some familiar movements: vultures, which have appeared out of an apparently empty sky, and are now drifting slowly down towards the ground on their broad wings. As the swallow approaches, he sees a pair of male lions astride the bloodied corpse of a zebra, with a dozen or so hyenas standing around like delinquent teenagers, impatiently waiting their turn. The vultures keep their distance too but, once the lions and hyenas have eaten their fill, they will tear greedily at what remains, performing their vital role as the clean-up squad of the African savannah.

The swallow is not a scavenger like the vultures: he hunts and kills for himself. Although he cannot yet hear the flies buzzing around the zebra carcass, his acute eyesight has picked out tiny, tell-tale movements that give away their presence.

A twist of his wing, and the swallow changes gear, flying lower than usual, the ground beneath him skimming past like a speeded-up film. Impala, a herd of elephants, a giraffe flash by. A lone human figure, walking along the riverbank, is soon left behind.

Wings swept back, the swallow dives, then swoops up towards his prey. A split second before impact, he opens his beak wide and strikes. Beak and fly make contact, beak snaps shut; the fly's world goes black. Another few milligrams of food; another burst of energy; another few hundred metres of fuel for the swallow's long journey north.

It's a journey that will take him many weeks. Today, he is just one of a loose flock of swallows following the course of the river that runs along South Africa's borders with Botswana and Zimbabwe, heading towards Mozambique and the Indian Ocean.

In a week or so, he will turn north towards the cradle of evolution, the Great Rift Valley, where Masai tribesmen sing and dance, and archaeologists discovered evidence of our earliest ancestors. Soon he heads north-west, crossing the world's greatest desert, the Sahara, until he reaches the sanctuary of a North African city, where the cry of the muezzin calls the faithful to prayer.

After seeing the sun set in Africa for the last time, he will watch it rise over the Mediterranean island of Mallorca, then a day or two later hear a chorus of nightingales in a Provençal vineyard. Finally, he will cross the Channel from Cap Gris Nez to the White Cliffs of Dover, almost home at last.

But all this lies far in the future. Our swallow cannot afford to look ahead. His attention must be focused on the present, for with the present comes constant danger.

High above, in the deep blue African sky, a falcon is circling. It's a hobby, a bird born and raised in a disused crows' nest in an English wood, only a few miles or so from the village where the swallow is heading. Just as the swallow saw the flies, so the hobby has seen the swallows. Wings back, eyes fixed on the target, she sweeps down towards the tiny blue-black specks below. The air whistles past her feathers: a new and unfamiliar note amid the usual late afternoon hum of insects.

At first, the swallow notices nothing wrong. Then, something deep in his consciousness warns of danger. He and the birds around him begin to speed up their flight; each is glancing right, left, above, below ...

Above – that's where the danger lies. There's a cry of alarm, but for one swallow it's already too late. In a couple of seconds, it's all over. The hobby stretches out her claws, brakes moment-arily, and grabs a small bundle of feathers. A journey which had only just begun is already over.

Our swallow has been lucky. He was not the hobby's target – not this time – though he did feel the rush of air and hear the thud of impact. It is a useful lesson: keep alert, keep aware, keep focused. If he fails to survive today, there will be no tomorrow.

There we leave him, high above the mighty Limpopo River as it glints in the late sun. Heading north, to a Somerset village so very far away. The journey has begun.

I

SWALLOWS

The Swallow is the glad prophet of the year – the harbinger of the best season: he lives a life of enjoyment amongst the loveliest forms of nature: winter is unknown to him; and he leaves the green meadows of England in autumn, for the myrtle and orange groves of Italy, and for the palms of Africa.

Sir Humphrey Davy,
Salmonia: Or, Days of Fly Fishing (1828)

I love swallows; they are one of my favourite birds. But that was not always the case. For the first forty-plus years of my life I lived in London, where the commonest member of the swallow family is the house martin, while the swift, whose flocks tear screaming across the city skyline, is the true sign of summer. Swallows were at best overlooked, and at worst simply irrelevant: I hardly even noticed the date when I saw the first returning bird of the year. These birds' annual reappearance made virtually no impact on me as a city-dweller, even though they had flown many thousands of miles to get here.

Then, well over a decade ago now, I moved to the countryside with my young family. We chose to settle in a Somerset village which, I soon discovered, boasted large numbers of sheep, a worrying lack of streetlights, and an abundance of swallows. On the day we moved in – 19 July 2006, then the hottest day of the hottest month on record – the swallow was the very first bird I saw flying over my garden. Soon afterwards, I made this note in my diary:

House sparrow and greenfinch are permanent fixtures, while sparrows also congregate on the bushes by the farmyard next door, and collared doves perch on the roofs of the buildings. Meanwhile, a constant procession of swallows and house martins (mainly swallows) passes overhead, usually heading westwards into the prevailing wind.

Having moved from city to country, it took me a while to shake off my rather abrupt urban manner, and adopt a friendlier, rural one. At first, I would be frustrated at having to wait in a queue in the village shop while the people behind the counter chatted to the customers. I also, to my subsequent embarrassment, asked our local farmer how his 'herd of sheep' was doing, and panicked the first time I had to cycle back from the local pub in the pitch dark.

But over time, I settled down to my new life. I can mark the very date when I realised I had 'gone native', and finally become a proper country-dweller.

It was 6 April 2010 – my fourth spring here in Somerset – and I had taken my children out for a walk along the lane behind our home. On our way back, I caught sight of an unfamiliar movement: a small, determined bird heading low over the field, forging a path into the sharp north-westerly breeze. It was, of course, a swallow – my first of the year. For the very first time in my life I felt that rush of excitement and emotion that marks this classic sign of spring, and, though the children were happily unaware of the moment's significance, I'm not ashamed to say I wiped away a tear.

I soon realised I shared this feeling with many others, not just in my village, but right across the Northern Hemisphere, as each year they too welcomed the swallows back. As the poet Ted Hughes so perfectly put it, 'The moment they dip in, and are suddenly there ...'

'You would have to be very dull of soul indeed not to be moved by the life of the swallow,' reflected another former Poet Laureate, Andrew Motion, in the BBC-TV series *Birds Britannia*. 'We time our seasons by its coming and going, in an absolutely primitive and very ancient kind of way – it's in our bones.' Veteran birder and TV presenter Tony Soper agreed. 'The way swallows come and whistle and sing is a very joyous arrival; and the fact that they make their home in your outbuildings is something you really look forward to each year.'

Having finally appreciated the swallow as a sign of spring – albeit rather late in life – I began to think about where I had seen the species before. I soon realised I had watched these birds in many parts of rural Britain, from Scilly to Shetland. Looking back over the thirty years of field notes from my birding trips abroad, I also noticed that wherever I had travelled, the swallow had been a more or less constant presence.

On my very first foreign birding holiday, to northern India in 1985, I had watched swallows skimming over the vast wetland of Bharatpur, and hawking for insects around the Taj Mahal. A decade later, on a filming trip to Florida, I'd come across huge flocks near Lake Kissimmee. I'd noted them on family holidays to Menorca and Mallorca, and while birding in Israel, Jordan, Mexico and Morocco. I'd seen them flying among elephants on game reserves in Kenya's Masai Mara and Tanzania; feeding above lions and wild dogs in the Okavango Delta of Botswana; over a sea lion colony along the coast of Patagonia; and even during our honeymoon in the Gambia. Swallows had, it seemed, followed me on my travels right around the world.

But despite all this documentary evidence, I could recall hardly any of these encounters. Perhaps I'd been more interested in the many new and exotic species of bird I saw, or transfixed by the other wildlife spectacles I was witnessing. My only excuse is that the swallow is such a ubiquitous bird I simply took it for granted.

There is one exception to my amnesia. It was June 2003, and I was filming in Iceland with the TV presenter Bill Oddie. On the final day of the shoot, we hitched a helicopter ride with the Icelandic Coast Guard to the volcanic island of Surtsey. Having erupted from the ocean floor in November 1963, this new land had the unusual distinction of being younger than any of us.

We only had a couple of hours on Surtsey, so we concentrated on filming the gull colony and various scenic shots, under a surprisingly warm sun. Then, literally out of the blue, a small bird appeared: a lone swallow, hawking for insects over our heads. Swallows are a very rare visitor to Iceland, the only European country where they do not breed. Seeing a swallow so far outside its normal range made me appreciate – perhaps for the first time – this species as a true pioneer.

Ever since David Lack published his classic monograph *The Life of the Robin* in 1943, there has been something compelling

Swal-low.

about following the life history of a single species of bird. Lack presented a familiar creature in a wholly new way, combining revelations about its day-to-day lifestyle with accounts of its interactions with human beings.

In the years following the Second World War, the publisher Collins produced, as a companion to its classic New Naturalist series, a series of short monographs. These included Stuart Smith's *Yellow Wagtail*, Guy Mountfort's *Hawfinch* and Edward Armstrong's *Wren*, which also delighted in both the behaviour of the species and its cultural significance. Angela Turner's *Barn Swallow* is a fine later example of the genre.

My own offerings – short biographies of the robin and the wren – are a more modest contribution to this long and distinguished tradition. Like previous authors, I have combined my own field observations with an account of the life cycle of the bird: its habits, distribution and range, breeding and migratory behaviour. I have also looked at the part each species plays in our traditional and popular culture; a subject I consider just as interesting and important as the science. As I noted in the introduction to the first book in this series, 'Alongside the real, biological robin, there's also a cultural and historical "robin".' With those species most familiar to us, these two identities are often inter-related, as the folklore is usually rooted in a particular aspect of the bird's behaviour.

When I came to write a third monograph, I was looking for another bird that not only led a fascinating lifestyle but was also deeply rooted in our consciousness, with a wealth of – perhaps unexpected – literary and cultural associations. But after two largely sedentary birds, the robin and the wren, this time I wanted to explore the story of a global traveller. I soon realised I need look no further than the swallow. The bird is a paradox: we see it on a daily basis for half the year, only for it then to disappear, and remain absent from our lives for six months or more, though living on in our collective memory.

As I probed more deeply into the swallow's extraordinary life cycle, I realised another paradox; one I, like, I suspect, many others, had overlooked. We consider the swallow to be one of 'our' birds – no less, indeed, than the robin or wren that haunt our gardens all year round. Yet unlike these species, we share the swallow with millions of other people across the globe. And while we think of our swallows 'spending the winter' in Africa, that is only half the story: they may indeed spend *our* winter in the Southern Hemisphere, but there they are welcomed as a sign of the coming of spring.

Indeed, the swallow is one of those few fortunate creatures that experiences a perpetual spring and summer (along with a touch of autumn) – never true winter. Some pay a price for this: that 12,000-mile round trip, which enables them to enjoy a life of more or less eternal sunshine, is full of unexpected hazards.

Until very recently, the greatest scientists and thinkers simply refused to accept that a bird weighing less than an ounce could possibly make such an epic, twice-yearly journey halfway across the planet and back – a voyage that has rightly been described as 'one of the wonders of the natural world'. The eighteenth-century Swedish scientist Carl Linnaeus, known as 'the Father of Taxonomy', confidently asserted that swallows spent the winter hibernating at the bottom of lakes, before emerging the following spring; a delusion also propounded by many of his contemporaries, including the great thinker Dr Samuel Johnson and the naturalist Gilbert White.

An earlier observer, Charles Morton, seriously suggested that swallows do indeed head away from our shores each autumn: not to Africa, but all the way to the Moon, a journey he calculated would take roughly four months, with the swallow flying at speeds of up to 125 miles per hour! Even the poet John Dryden gave some credence to this bizarre explanation for migratory birds' absence:

> They try their fluttering wings, and thrust themselves in air,
> But whether upward to the moon they go,
> Or dream the winter out in caves below,
> Or hawk at flies elsewhere, concerns us not to know.

The truth is, if anything, even more incredible. That this bird, which hatched out of a tiny egg just a few weeks earlier, can, without any help or guidance from its parents, find its way to the southernmost part of the vast African continent – and all the way back the following spring – does indeed seem, in the original meaning of the word, 'not believable'. Yet that is exactly what swallows do.

What follows is, therefore, very different from the story of the robin or the wren. Though their lives have plenty of drama, they are essentially parochial and domestic. In the swallow's story, both the bird and I travel halfway across the world and back. To whet your appetite, let me repeat the quotation that serves as this book's epigraph, from the eminent twentieth-century ornithologist Collingwood Ingram: 'My little book is concerned with the life history of a bird which is beyond doubt the best known, and certainly the best loved, species in the world – the swallow.'

When I first came across this pronouncement, I was, I must say, rather taken aback. Was it *really*? But having since spent a year watching swallows in Britain and South Africa, reading about their travels, and talking to friends in other countries who are also visited by this globetrotting species, I have come to believe that Ingram was absolutely right.

We are so accustomed to swallows that sometimes it is hard to appreciate just how unusual they are; especially if we compare them with other songbird families in the *Passeriformes*, the order which includes more than half of all the world's bird species. Thrushes and tits, robins and sparrows, flycatchers and finches, and larks and pipits: all may vary somewhat in shape, size, colour and pattern, but essentially, they all fit the same basic template. That's mainly because they obtain their food either from the ground or while foraging among vegetation, rather than in the air.

Swallows, in contrast, are effectively flying machines, whose entire body shape and structure has evolved to make them superbly efficient at what they need to do: catch airborne insects while on the wing. All members of the swallow family have long, pointed wings and slender, streamlined bodies; some also have long tails, with streamers on the outer sides, which they can use as both rudder and brake to turn or slow down. These features allow them to fly more efficiently, and be far more manoeuvrable, than other birds, and snatch a fast-flying insect in mid-air. Their unusual shape also lowers their metabolic rate, so they can stay airborne, and hunt, for longer than other birds of the same size. Their long, pointed wings are a huge asset during migration, producing less drag to enable the swallow to fly further and faster, while expending far less energy than birds with shorter, more rounded wings.

Swallows' eyes are more forward-facing than those of most other songbirds: more like a raptor's – falcons and hawks also depend on catching prey in high-speed, aerobatic, flight. They

SWALLOWS

HOUSE-MARTINS

have two separate foveae in each eye (the depression in the retina which provides the highest visual acuity), so they can see detail to either side as well as in front of them. A final adaptation gives them short bills with a very wide gape, ideal for grabbing their aerial prey.

Swallows and martins (the latter name usually refers to the smaller, square-tailed species) originated at least 50 million years ago in either Africa or Asia; over time, they spread right around the globe, evolving into about eighty different species. Collectively known as 'hirundines', from the Latin, they are one of the world's most successful bird families, and owe that success largely to their diet. Flying insects are found all over the world, in more or less every habitat apart from the highest mountain ranges and the polar regions. As a result of the omnipresence of their prey, swallows and martins are pretty much ubiquitous too: they fly and feed above fields and meadows, woodland and scrub, coasts and marshes, lakes and rivers and, in many parts of the world, over our towns and cities.

Today, swallows and martins can be found on six of the world's continents, while one plucky barn swallow has even made landfall in the seventh, Antarctica. Two species, the Mascarene martin and Pacific swallow, breed on remote islands in the Indian and Pacific Oceans, while the Australian welcome swallow has, during the past century or so, managed to colonise New Zealand. None of the other 240 or so families of birds is so globally widespread.

The different swallow species do vary in size and weight. The smallest and lightest, such as the white-thighed swallow of South America and the fairy martin of Australia, are a mere 11 cm long, and weigh as little as 9 or 10 grams. At the other end of the scale, the mosque swallow of Equatorial Africa can reach a length of 24 cm, tipping the scales at over 50 grams, while the purple martin of the Americas is shorter but even heavier, reaching a whopping 64 grams. The barn swallow is slightly larger than the average for its family: roughly 18 cm long, and weighing between 16 and 24 grams – about half the weight of a standard packet of crisps.

The various species also vary hugely in abundance and rarity. The barn swallow is by far the most widespread, being one of only two species that breed in both the Old and New Worlds (the other being the sand martin, known in North America as the bank swallow from its habit of nesting in sandbanks). As Collingwood Ingram noted: 'During its nesting season the [barn swallow] literally girdles the Northern Hemisphere and, what is more, it manages, by some miraculous means, to evenly distribute its entire population over that vast area.'

Several species have a minuscule range, covering just a few hundred square miles. These include the Tumbes swallow, found along the coasts of south-west Ecuador and north-west Peru; the Galapagos swallow, from the eponymous archipelago; the Congo sand martin, which mostly lives along the river of that name; and the mountain saw-wing, which haunts the forests and clearings of Mount Cameroon and the island of Bioko (also known as Fernando Po) in West Africa.

Despite their very restricted home ranges, however, none of these species is yet considered globally threatened. The same cannot be said of several other members of the family. The blue swallow of southern and eastern Africa (See Chapter 5), the white-tailed swallow of Ethiopia and the Bahamas and the golden swallow of the Caribbean are all classified as 'Vulnerable' by BirdLife International, because of their small and declining populations.

Yet two species are rarer still. The white-eyed river martin was discovered at a lake in central Thailand as recently as 1968. Despite its very distinctive appearance – glossy-black with long tail streamers, a white band across its rump and the prominent white eye-patch that gives the species its name – it has hardly ever been seen since. Several specimens were collected from roosts over the next few years, and there have been a handful of reported sightings, but none have been seen at all since 1980, despite thorough surveys of the area and neighbouring regions. There seems little hope that the white-eyed river martin will ever be seen again.

The Red Sea swallow is even more of a mystery – for it may never have been seen alive at all. One dead specimen was obtained from a lighthouse north of Port Sudan in May 1984, yet since then there have only been a handful of inconclusive and unproven sightings elsewhere in the Horn of Africa. Aptly, the scientific name of this enigmatic species, *Petrochelidon perdita*, means 'lost rock swallow'.

Despite sharing that same basic body shape, which makes them so aerodynamic, not all swallows and martins migrate. Indeed, some, including the rarest and most range-restricted species, are highly sedentary. Yet those species we are most familiar with in the Northern Hemisphere – the barn swallow, house and sand martins of the Old World, and the purple martin, tree and cliff (and bank and barn swallows) of the New – are all long-distance migrants. Each has evolved to undertake epic migrations to and from our more northerly latitudes, to breed here in the boreal spring and summer, and take advantage of the abundance of flying insects and reduced competition from rival species.

*

Of all these the barn swallow is, without question, the greatest global traveller. It is the only species to use all three major migratory routes: the African-Eurasian, East Asian-Australasian and Pan-American flyways. Its world range extends over an astonishing 20 million square miles (nearly 52 million square kilometres) – considerably larger than Asia, and more than five times the size of Europe. We all know that swallows migrate south to Africa in autumn and back to Britain in spring. Yet others continue to travel way beyond the Arctic Circle to Varangerfjord in Norway, almost the northernmost tip of mainland Europe, and more than 70 degrees north of the Equator.

Swallows breed across virtually the whole of Europe and Asia, from Ireland in the west, almost as far as the Bering Strait in the east, and south to the southernmost point of China, Hainan Island – less than 20 degrees north of the Equator. These birds overwinter across a broad swathe of the Old World: western populations mostly migrate to sub-Saharan Africa, while more easterly breeders head down to the Indian sub-continent, south-east Asia and even northern Australia.

Of the swallows found in the Northern Hemisphere, British and western European birds travel the furthest, with the vast majority heading all the way to southern Africa – Botswana, Namibia and South Africa itself – a distance of at least 6,000 miles (almost 10,000 kilometres). Bear in mind that this is the most direct route: swallows will not fly in an absolutely straight line, and also need to stop and feed on the way, so it is likely that they actually cover well over 10,000 miles – and perhaps more – on each leg of their two-way journey.

In North America, barn swallows breed in Canada, Mexico, all forty-eight lower states of the USA, and south-east Alaska (and are a rare vagrant to the fiftieth state, Hawaii). Like other birds, they often take the shorter sea route across the western Atlantic, flying over the open ocean to their winter quarters in Central and South America, with a hardy few travelling as far as Tierra del Fuego, some 55 degrees south of the Equator. That they are so widespread is in some ways surprising: after all, until two or three centuries ago, the human population of North America would have been very thinly spread; only with the winning of the 'Wild West', and the advent of extensive cattle farming, would swallows have had the opportunity to colonise much of the continent.

In recent years, swallows have started breeding in parts of their winter range, including the area around Buenos Aires in Argentina and the mountains of northern Thailand, but not, as yet, in southern Africa – though it may only be a matter of time. Closer to home, swallows breed across virtually the whole of the British Isles, being absent only from a few offshore islands, parts of the Scottish Highlands and our inner cities. They are most abundant in Ireland, eastern Scotland, west Wales and south-west England, including my home county of Somerset. Here, during the spring and summer months, swallows are found everywhere apart from the high tops of Exmoor and the major town centres – where, as is the case throughout urban Britain, they are replaced by the house martin.

For those of us who live in the British countryside, the swallow, as the twentieth-century ornithologist David Bannerman noted, 'is so universally known that little needs to be written in the way of description. Its graceful flight, narrow wings and forked tail with elongated outer tail feathers enable it to be recognised immediately.' Yet people often tell me they find it hard to distinguish the swallow from its smaller relatives, the house and sand martins, and sometimes even from another (totally unrelated) species which spends its life in our summer skies: the swift.

We birders are not always known for our tolerance of those who don't share our all-consuming passion, yet I still find it strange that people who seem perfectly capable of telling apart the various species of tit and finch that visit their bird feeders, or can distinguish a coot from a moorhen in their local park, appear to have such a blind spot when it comes to this quartet of quite distinctive species. Perhaps it is simply because swallows, martins and swifts never seem to stay still, so are harder to get a good look at.

Back in 1908, the popular author and naturalist William Percival Westell was scathing in his dismissal of those unable to identify these common birds: 'How many people are there living in the country who know a Swallow from a Martin? To most people the Swallow, the Martin and the Swift are all Swallows … But who, may I ask, once knowing the bird, can possibly mistake it?' Unfortunately, Westell's contempt for the ignorant masses is rather undermined by his own ignorance about the swallow's winter quarters, as he suggests that they 'arrive from India and Ethiopia', rather than southern Africa (though, to be fair, he was writing before the first ringed swallow was recovered there).

So how different are swallows, martins and swifts from one another? By far the easiest to identify is the swift, which is about as closely related to swallows and martins as owls are to kestrels – that is, not at all. Their superficially similar appearance is via the process of convergent evolution, because both swifts and swallows hunt flying insects.

Swifts are not passerines at all, but in a completely different order, the Apodiformes. This also includes the hummingbirds,

4 – HEADS OF SWALLOW, HOUSE-MARTIN, SAND-MARTIN AND SWIFT

with which they share many physical characteristics and habits. Swifts only come to land when they visit their nests, and even then they struggle, as their tiny feet are set well back on their bodies to allow them to live an almost entirely aerial existence: the swift's scientific name, *Apus apus*, means 'no foot, no foot'!

Swifts look completely different from swallows: they are almost entirely sooty-brown (appearing black in most lights), and have long, narrow, curved wings shaped like a scythe; not at all like the triangular wings of the swallows and martins, which have a broad base and pointed tip. And while swallows and martins twitter, swifts *scream*, often rushing across the firmament like jet fighters, a unique contribution to the soundscape of the city skyline.

The two species we call martins – the sand and house varieties – are, to be fair, superficially similar in shape and habits to the swallow, though both are noticeably smaller and less elongated. They too have distinctive plumage features and that indefinable quality birders call 'jizz', so it doesn't take much practice to tell the three species apart.

The sand martin helps us by coming back from Africa a couple of weeks earlier than its larger cousins: usually by the middle of March. On blustery days, if I head up to Cheddar Reservoir in the lee of the Mendip Hills, I often find dozens of these new arrivals flying low over the choppy waves, picking up tiny insects. They have a pleasingly compact shape: shorter wings than the swallow, and a much shorter, square-ended tail. But the best way to tell the two apart is that sand martins are brown above and white below, with a brown band running across their upper breast, beneath a white throat.

Like all members of their family, sand martins are sociable birds. A couple of years ago, on a bright, sunny morning in late April, I was sitting in the Tor View Hide at the RSPB's Ham Wall reserve in Somerset, when there was a burst of activity in the adjacent reedbed, accompanied by a tuneful chirping. The only other person in the hide – a photographer laden with lenses and camouflage gear – suggested this racket might be starlings; but the date was wrong and, besides, it didn't sound quite right.

Then, without warning, hundreds of sand martins rose as one from the reeds and began to hunt for insects. For the next half hour or so, they swooped back and forth, uttering their chirruping calls before, as quickly as they had come, zooming away – probably northwards to their nesting colony in a bank of a river or gravel pit. Afterwards, I wondered if I had really witnessed it.

The house martin, as the name suggests, is even more closely linked with our own lives. The French call it the *hirondelle de fenêtre*, or 'window-swallow', which, along with 'eaves-swallow', is also an English folk name for the species. Again, given close views (which you usually get), house martins are easy to identify. They are blue-black above, white below, and with a prominent white rump above a dark, short and slightly forked tail (whereas the swallow's upperparts are completely blue-black). Overall, as Bill Oddie has pointed out, their colour and pattern bear a striking resemblance to a rather larger predator, the killer whale.

House martins also nest on the outside of buildings, rather than inside, as most barn swallows do.

Sadly, two of these three species are not doing very well at the moment. Swift numbers are down by more than half in the past twenty years or so, and they are now on the Amber List of Birds of Conservation Concern. The fall is due mainly to the lack of suitable nest spaces, which disappear when old buildings are repaired or restored, while new buildings often fail to provide suitable gaps where the birds can nest. Now conservationists are encouraging individual homeowners and corporate housebuilders to install 'swift bricks', which have a hollow compartment where the birds can nest, but even this may not be enough to reverse the species' decline.

If anything, the plight of the house martin is even worse, with a serious and rapid fall in numbers in many towns and cities during the past two decades. However, perhaps because of the short-term benefits of climate change, Scotland seems to be bucking the downward trend – for the moment at least. Like the swift, the house martin is now also on the Amber List. Astonishingly, we do not even know exactly where British and European house martins spend the winter – beyond 'sub-Saharan Africa' – because, unlike barn swallows, they do not roost communally.

Sand martin numbers have always fluctuated, with a major crash in the late 1960s and early 1970s following a drought in the Sahel Zone just south of the Sahara Desert, en route to their West African winter quarters. Although numbers did bounce back, and are currently on the rise, like all long-distance migrants the sand martin is very vulnerable to sudden environmental change.

And what of the swallow itself? The jury is still out, but the figures do seem to show that, following a rapid rise in the population of breeding birds towards the end of the last millennium, numbers have in the past few years begun to fall again. So, we cannot afford to be at all complacent about the fate of this special bird.

Like so many of the words we use for our commonest and most familiar birds, the origins of the name 'swallow' are rather vague. The eminent philologist W. B. Lockwood, author of *The Oxford Book of British Bird Names*, suggested that it derives from a hypothetical Proto-Germanic word (which would be several thousand years old at least) meaning 'cleft stick' – a reference to the bird's forked tail.

What we do know is that the swallow is one of just sixteen species of bird whose names occur in Anglo-Saxon literature. In one of the earliest collections of Old English writings, the Épinal-Erfurt Glossary (two manuscripts compiled for Aldhelm, the Abbot of Malmesbury, around the year AD 700), it appears alongside the Latin word *Hirundo* as '*sualuuae*'.

After the name entered the English language, it was applied to several other creatures that also sport forked tails or similarly protruding appendages. Another graceful and elegant family of birds – the terns – are widely known as 'sea swallows', a folk name that also occurs in other Germanic languages including Dutch and German. The scientific name for the common tern,

Sterna hirundo, repeats this theme, as does that of the world's largest member of the family, the Caspian tern, *Hydroprogne caspia*, which translates as 'water swallow'. One large family of butterflies – the swallowtails – and a species of moth, the swallow-tailed, also carry the epithet.

Like other common birds, the barn swallow has acquired a range of folk names – though not perhaps as many as we might imagine, considering it is such a familiar bird. Gilbert White preferred 'chimney swallow', while 'house swallow' was also occasionally used by early naturalists, who translated it from a German folk name for the species, *Haus-Schwalbe*. This never gained wide currency, probably because of potential confusion with the house martin, but is still used for the Javan subspecies of the Pacific swallow.

Other folk names include 'red-fronted swallow', from the bird's rusty throat; the self-explanatory 'forktail'; 'tsi-kuk' (presumably based on the bird's alarm call) from Cornwall; and a series of Scots names: 'swallae', 'swallock' and, my favourite, 'witchags', which means 'little witches', and is still occasionally heard today.

Perhaps because they are such ubiquitous birds – for half the year at least – swallows have held a central place in the cultures of the Northern Hemisphere for many millennia. In the fourth century BC, the Greek philosopher and writer Aristotle famously cautioned that 'One swallow does not make a summer': a warning against assuming that a single event stands for a wider trend. Swallows appear even earlier than this, as hieroglyphs on Ancient Egyptian tombs, in which they signify both the bird itself and other, more symbolic meanings.

In Ancient Egypt, swallows were often associated with the souls of the departed, and mummified specimens were placed in tombs. In this extract from the Book of the Dead, a series of scrolls containing prayers and spells dating back to roughly 1500 BC, swallows are described as guiding the departed spirit through the underworld, and ultimately helping them to achieve eternal life: 'I have gone to the great island in the midst of the field of offerings, on which the swallow gods alight; the swallows are the imperishable stars.' Perhaps it was the effortless ease with which they fly that led swallows to be regarded as suitable pilots for this final journey.

The Ancient Egyptians also associated swallows with the coming of a new dawn, probably because they are among the first birds to appear in the skies after sunrise. However, as Angela Turner, who has written extensively about both the biology and culture of swallows, has pointed out, the Egyptian population of barn swallows is, unusually for this species, resident all year round. That explains why, instead of being linked with a seasonal phenomenon (the coming of spring), the bird is more closely associated with a regular diurnal one (the sunrise).

In medieval heraldry, the swallow's close relative the house martin (known as the martlet) represented younger (usually fourth) sons. Because the firstborn sons traditionally inherited their father's entire estate, and the other elder brothers went into the church, the youngest were doomed to wander the globe in search of a livelihood and fortune. This connection with wandering has led some to suggest that the martlet is actually the swift, whose lack of obvious feet makes it a prime candidate for

such a symbol. However, every depiction of the martlet – including that on the shield of the English county of Sussex – looks far more like a swallow or house martin than the more attenuated, scythe-winged swift.

Further afield, swallows are the national bird of Austria and Estonia; but oddly, not a single species of swallow or martin is the official bird of any US state or Canadian province, despite the purple martin being one of the North American birds most closely associated with human beings, who provide communal nestboxes for these annual visitors.

Swallows also feature strongly in literary and popular culture. In the final line of one of the most quoted poems in the English language, 'Ode to Autumn', the Romantic poet John Keats writes that 'gathering swallows twitter in the skies'; while their sound is also featured in Homer's epic poem *The Odyssey*, dating back to the eighth century BC. The twanging of bow-strings is compared to the twittering of swallows – most likely a reference to the swallow always returning to the same place at the end of its prodigious voyage – just as Odysseus eventually does.

A more recent reference to the species appears in the title of one of the twentieth century's best-loved children's books, Arthur Ransome's *Swallows and Amazons*, published in 1930. The story was partly based on the Altounyans, a real-life family of Anglo-Armenian origin, who had shared a summer holiday in the Lake District with Ransome two years earlier; but was also inspired by the author's own childhood experiences. The 'Swallow' in the title is the name of one of two dinghies in which the Altounyan children (renamed the 'Walker family' for a domestic audience) learn to sail; but Ransome also used 'Swallows' as a collective name for the children themselves, in their rivalry with another gang (the 'Amazons').

There is a further nautical association: a swallow in flight has long been a popular design for a tattoo, especially among sailors. Traditionally, a sailor would have a tattoo of a single swallow when he had travelled more than 5,000 nautical miles (5,750 miles or 9,260 kilometres), and would add a second once he had reached the milestone of 10,000 nautical miles (11,500 miles or 18,520 kilometres). According to the online Urban Dictionary, if a sailor should drown, swallows will fly down and lift his soul up to heaven – just like the Ancient Egyptian belief. It goes on to note that boxers also favour the swallow tattoo, as for them it

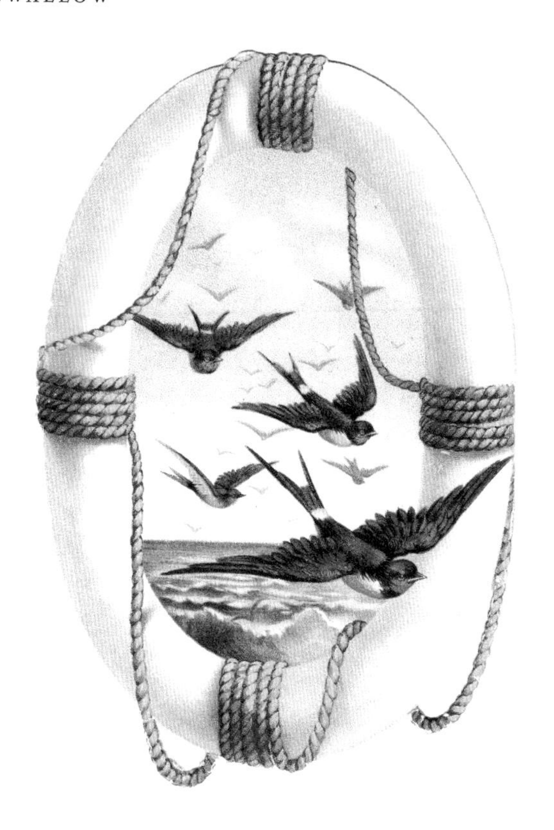

symbolises speed and power; tattooed across the knuckles, the swallow warns an opponent that 'these fists fly'.

It is tempting to suppose that the swallow was chosen by sailors because, as with the heraldic symbol of the martlet, these birds represented the wandering spirit; yet the opposite may actually be the case. Swallows return to the place where they were born each year, so it is more likely that, in an echo of *The Odyssey*, sailors chose the bird as a reminder of home; a kind of guardian angel, ensuring a safe homecoming to their loved ones.

This annual return, each spring, is by far the most resonant of the swallow's cultural associations. Now that the sound of the cuckoo is a fading memory for many of us, especially in southern Britain, where an entire year can pass without us hearing that familiar double note, the appearance of the first swallow – a sign that the world has 'come back' to us in all its growth and fecundity – has become even more significant.

2

SPRING

I wonder if the swallows that nest in the chimney of my Suffolk farmhouse have the faintest idea how profoundly they affect my emotions. When they first arrive from the south in spring, and I hear the thrumming of their wingbeats amplified to a boom by the hollow brickwork, my heart leaps.

Roger Deakin, 'Follow the Swallows'
(posthumously published, 2015)

On a clear day, from the French coast between Calais and Boulogne, you can sometimes see the White Cliffs of Dover. But this is not a clear day. Mist shrouds the headland of Cap Gris Nez, with a fine drizzle adding to the misery of those hardy, or reckless, enough to venture outdoors.

For our swallow, even fine rain is bad news, seeping into his feathers and keeping the few early-spring insects in their hiding-places. He is on the last leg of a journey all the way from the tip of Africa, heading for a barn in Somerset, where he will find a mate, build a nest, and raise his young. But for now, he must find shelter from the rain.

A couple of hours later, the mist eventually lifts, to reveal a fine, clear spring morning. Embarking on this short sea crossing means being constantly aware of danger. The English Channel is the busiest shipping lane in the world and, because he flies so low over the water, he must take care not to be sucked into the slipstream of a passenger ferry.

An hour later, having beaten the boat, he has reached our shores. Today, there are not bluebirds but swallows over the White Cliffs of Dover. At a fairground below, Easter bank holiday crowds are queuing up for the helter-skelter and merry-go-round, licking toffee apples and candy floss. No one notices the small bird passing a few feet above their heads.

Another hour, and our swallow reaches the start of the suburban sprawl cast like a crumpled tea-towel over most of south-east England. Here, insects are few and far between; the fumes from the city and its traffic have seen to that.

The swallow is getting hungry, so he keeps one eye out for the tell-tale glint of water. Ignoring the rivers, streams and occasional open-air swimming pool, he spies a familiar landmark ahead. A broken line of silver-grey: a chain of gravel pits and reservoirs, dug after the Second World War to provide building material for houses and roads, and water supplies for a growing city. Not, perhaps, the most beautiful prospect, but for this bird, a very welcome sight – for water attracts insects, and insects mean food.

The swallow is not the only migrant using these artificial waterways as a service station. It's still a few weeks too early for swifts, but little flocks of house martins and sand martins are already feeding on the insects that gather above the surface of the water. The swallow joins them, opening his beak wide and then snapping it shut around each unwary morsel.

Tempted by the sunshine, a birder has dug out her binoculars and cycled the mile or so from her home to the reservoirs, hoping to catch up with these early spring migrants. As she scans the reservoir, the sun catches the lenses of her binoculars and temporarily dazzles the swallow; it swerves quickly away. As the bird turns, his dark-blue upperparts contrast with a creamy-white belly, red forehead and throat.

It is this birder's first swallow of the year, bringing joy to her heart, as other returning swallows are now bringing delight to millions of people, all the way across the Northern Hemisphere. Finding a dry patch of grass, she lies back and stares up at the cloudless sky. One swallow may not quite make a summer, but it is still a sign, and a good one at that. Spring is definitely here.

But our swallow has no time to rest. He is driven by instinct, and that instinct is urging him on. So, veering upwards from the reservoir, he heads west: away from London, passing over Heathrow Airport, across the M25, and into a landscape of fields and woods, hedgerows and streams; out into the countryside, on the last leg of the long journey home.

A few hours later, he is only a few miles short of his target, out of the 6,000 he has already flown. During this final stage, he uses visual clues to find his way, following the lines of rivers and hills, adjusting his direction as he begins to notice familiar landmarks.

The smog and fumes of the capital are a distant memory. He flies through cool, clean air, over the patchwork quilt of fields, bounded by hedgerows, that makes up so much of lowland Britain. It could hardly be more different from the African savannah he left just a few weeks ago: criss-crossed by broad rivers and dotted with baobab trees, where elephants, giraffes and lions roam under vast skies.

From Alaska in the west to Japan in the east, swallows are arriving back to breed. Well over one million in Britain alone; perhaps thirty million in Europe, and tens of millions more across Asia and North America. Everywhere they go their presence is unconsciously but gladly welcomed, as it brings sunshine and warmth to mark the end of a long, cold winter.

Half a mile ... a few hundred yards ... a hundred feet ... As he approaches the farmyard, a landscape he recognises is laid out below: a tractor corrugating the ground, someone threading scraps of colour onto a washing line, a flock of sheep being chivvied into a field by the farmer's border collie.

In a week or so's time, another swallow – a female – will also arrive back. After a brief but intense courtship, the pair will build a nest on the rafters in the barn; a nest made from mud, grass and hair, lined with a few feathers shed by the farmyard chickens. Here, the female will lay five smooth, oval-shaped eggs, off-white with reddish-brown spots. And if this first clutch is successful, she and her mate will have a second, and perhaps even a third, brood.

But all that is far into the future. As the bird swoops down, the farmer looks up from his ploughing. He has heard the swallow's light-hearted twittering, and it takes him a moment or two to realise where it comes from, before a broad smile spreads across his face.

Why is the swallow here at all? After all, this species – along with the martins, swifts and many other summer visitors to Europe, Asia and North America – originally evolved in Africa; they are birds of the Southern Hemisphere that somehow found a temporary home in the north. How they did so is a mystery; but we can infer that it would have occurred over many thousands of years, in short stages rather than one giant leap.

Let's start with the fact that many species in Africa undertake what is known as 'intra-African migration': making seasonal movements between their breeding and non-breeding grounds, to take advantage of the wet and dry seasons occurring at different times in different parts of the continent. Being dependent on flying insects, and able to fly long distances, it is likely that the barn swallow was one of those species that headed north, and over time went further and further, eventually reaching what we now call the Sahara Desert.

We might imagine that they would have stopped there – after all, what benefit could be gained from heading into this arid emptiness? But that ignores the fact that for roughly nine millennia, from 14,600 to 5,500 years ago, the Sahara was actually quite fertile, with areas of grasses, trees and lakes rather than the vast sands we know today. Known as the African Humid Period, and the result of temporary changes in the Earth's orbit around the sun, this would have enabled swallows to continue northwards to reach the Mediterranean and ultimately cross over to Europe.

Their timing was perfect: the last of the Ice Ages, which had brought freezing conditions across much of the continent, had just come to an end, so, over a number of generations, swallows – and other migrant birds – could continue to forge northwards, finding food in abundance, thanks to the long hours of daylight during the boreal summer.

'Of all the families of birds which resort to this island for incubation, food, or shelter', declared the late-eighteenth-century ornithologist and publisher Thomas Bewick,

> there is none which has occasioned so many conjectures respecting its appearance and departure as the Swallow tribe ... Their arrival has ever been associated in our minds with the idea of spring; and till the time of their departure they seem continually before our eyes.

'Their willingness to nest and forage close to people', adds Angela Turner, 'has made them a part of the fabric of farmyard and village life.'

Wherever they live and breed, swallows have been integral to rural life for longer than we might think. The archaeologist Dale Serjeantson, of the University of Southampton, has suggested that our cave-dwelling ancestors may well have shared their homes with nesting swallows, and looked out for them returning each year – as spring would have been such a crucial season in their calendar. Later, during the seventh or eighth century BC, the Greek poet Hesiod (a contemporary of Homer) was the first to write of the swallow as a marker of the new season.

The well-known proverb 'One swallow doesn't make a summer' (and its several variations) is first noted in the writings of the Greek philosopher Aristotle, during the fourth century BC. But it is undoubtedly far older; the same sentiment appears, roughly 250 years earlier, in *Aesop's Fables*.

'The Young Man and the Swallow' (sometimes known as 'The Spendthrift and the Swallow') tells of a callow youth who has wasted all his money on luxuries and gambling, until he only has one valued possession left: a cloak. When he sees the first swallow of the year, he decides that spring has arrived, and he no longer has any need of warm clothing. But no sooner has he sold his cloak than weather turns cold again, and he finds the swallow frozen to death. Soon afterwards, he too dies of cold. The moral is clear: just as the arrival of the first swallow is but a harbinger of spring – not a sign that there will be no more wintry weather – so we should not draw broad conclusions from a single event.

In many cultures, a specific day marks the timing of the birds' return, as it does with another sign of spring, the cuckoo. The Russians believed that swallows flew in from Paradise on 25 March each year, the same date chosen by French and German folklore. This may be because it is a key event in the Christian calendar: the Feast of the Annunciation, marking the visit of the

angel Gabriel to the Virgin Mary, to tell her that she would be giving birth to the Son of God.

Because the arrival of swallows tends to roughly coincide with Easter, an inevitable connection arose between the bird and the religious festival – helped, perhaps, by the swallow itself being regarded as a symbol of natural rebirth. Like other birds, swallows are often linked with the crucifixion and resurrection of Christ: the red on the swallow's forehead and throat is supposed to have come about, like the goldfinch's crimson face, from trying to remove the Crown of Thorns; or, as some believe, from the bird trying to pull out the nails from Christ's hands and feet.

The religious link between humans and swallows may go back even further, according to one legend chronicled by the Victorian folklorist Charles Swainson. When Adam was bewailing his fate, having fallen from Paradise to Earth, a swallow is said to have come to him, taken a strand of his hair to Eve, and intertwined it with hers. Thus were the couple reconciled, and for this good deed, the swallow was 'allowed to nestle in the dwellings of men'.

In the German province of Saxony, swallows are supposed to arrive on Palm Sunday – rather odd since, like Easter, this is a variable date which can occur anytime between late March and the middle of April. Gilbert White recalled that when he was a boy, he 'observed a swallow for a whole day together on a sunny warm Shrove Tuesday; which day could not fall out later than the middle of March, and often happened early in February'. Sadly, he was unable to recall the exact year, so we cannot ascertain just how unusual a record this was.

In south-east Europe, notably Bulgaria, Macedonia and Greece, the return of the swallow has given rise to a longstanding tradition. From Baba Marta (Grandma March) Day – 1 March, the first official day of spring – local people wear a *martenitza*, a pair of knitted dolls, made from red and white wool, until the wearer sees either a swallow, a stork, or a tree in blossom, signifying that spring has really arrived.

Also in Greece, on the island of Rhodes, groups of boys used to visit houses around their town or village on the day of the birds' arrival, singing a 'swallow song' and asking for food, in a custom known as the *chelidonia* (from the Greek word for swallow, *chelidon*):

> The swallow, the swallow is come
> Bringing good seasons and a joyful time ...

Closer to home, poets and writers down the ages have celebrated the swallows' return. In his poem 'What the Farmer's Wife Said', Ted Hughes talks of their arrival, after months during which we have hardly given the absent birds a second thought, being 'the loveliest thing' about them; while Christina Rossetti could hardly contain her excitement when anticipating their annual return:

> It's a year almost that I have not seen her:
> Oh, last summer green things were greener,
> Brambles fewer, the blue sky bluer.
>
> It's surely summer, for there's a swallow:
> Come one swallow, his mate will follow,
> The bird race quicken and wheel and thicken.
>
> Oh happy swallow whose mate will follow
> O'er height, o'er hollow! I'd be a swallow,
> To build this weather one nest together.

One of the reasons we extend such a warm welcome to the returning swallows – as opposed to much commoner but less visible migrants such as the willow warbler, which outnumbers the swallow by three to one – is, as the Victorian nature writer W. H. Hudson pointed out, because they live so closely alongside us:

> When our summer migrants … seek their customary homes in woods and groves by the sides of streams and marshes, and on downs and waste lands, the swallow alone comes direct to us to deliver the good message, so that even the sick and aged and infirm, who can no longer leave their beds or rooms, are able to hear it. What wonder that we cherish a greater affection for, and are more intimate with, the swallow than with our other feathered fellow-creatures!

In a radio script published as an essay after his untimely death, the pioneering nature writer Roger Deakin also wrote of his joy at their annual return to his home in Suffolk:

> They seem to bless the house with the spirit of the south; the promise of summer. Swallows have such a strong homing instinct that it is quite possible this same family of birds, by now an ancient dynasty, has been returning here to nest for the 450 summers since the chimney was built.

Deakin's reference to chimneys reminds us that the bird was once known as the 'chimney swallow', the name preferred by

Gilbert White. Even today, the French often refer to the species as *Hirondelle de cheminée*, while the official name in German is *Rauchschwalbe*, meaning 'smoke swallow'. Although we may find it odd that these birds would choose to nest in chimneys, Angela Turner points out that at the time when these folk names originated, chimneys would have been larger structures than modern ones, with plenty of shelves where the swallows could build their nests.

In his *Natural History Letters*, the early nineteenth-century poet John Clare noted that:

> [The Chimney Swallows] come about the middle of April & I observe on their first visit that they follow the course of brooks & rivers I have observed this for years & always found them invariably pursuing their first flights up the brinks of the meadow streams.

New arrivals still seek out water – they are often seen feeding low over the surface of lakes, reservoirs and gravel pits, where insects congregate in the greatest numbers even on chilly, unsettled April days.

In some springs, if there is a spell of bad weather, returning swallows are often delayed. 'Last month we had such a series of cold, turbulent weather,' wrote Gilbert White in May 1770,

> such a constant succession of frost, and snow, and hail, and tempest, that the regular migration of the summer birds was much interrupted. Some did not show themselves till weeks after their usual time ... two swallows discovered themselves as long ago as the eleventh of April, in frost and snow; but they withdrew quickly, and were not visible again for many days.'

Like all migrants, swallows are also sometimes forced to retreat in the face of harsh and unseasonable weather. But the instinct to breed is a powerful one and, as soon as conditions take a turn for the better, they come back to the place where they hatched out a year or more earlier, to establish their territory.

First to return are invariably the males, often the older and more experienced birds which have made the journey from Africa before. If more than one swallow is occupying the same space, there may be skirmishes as each bird tries to defend the few square yards where he intends to build (or rebuild) his nest.

A few days later, the first females arrive, prompting the males into a frenzy of courtship: singing, like other songbirds, but also displaying to their potential mates, by spreading their tails out to reveal the bright white spots along the upper- and undersides. This is sometimes done as a group, the males rising into the spring sky to utter their curious warbling song, while flying around together.

W. H. Hudson described the swallow's song as 'more free and spontaneous than that of any bird ... the notes leaping out with a heartfelt joyousness which is quite irresistible'. Yet we don't usually think of it, unlike that of other familiar British birds like the robin or song thrush, as especially beautiful or tuneful. But to Collingwood Ingram, it was the *spirit* of it that really mattered:

> In my opinion no bird's song is quite so expressive of *la joie de vivre* as that of a swallow. To listen to its hurried, exuberant twitterings, that ineffably cheerful sound which seems as though the little bird's heart were on the point of bursting with happiness, is always a sheer delight.

Ingram also revealed that the French refer to the swallow's song as *le gazouillement*, the same word they use for a babbling brook. The nature writer Mark Cocker has felicitously described it as an 'aqueous burble ... a product of the surplus insect protein, converted through the birds' digestive system into the music of swift and swallow', or more succinctly, 'insect music'.

Yet, given that a typical burst of song lasts only a few seconds, it does not always register with us. I usually hear the twitter of groups of swallows above my head on fine spring or summer afternoons, when I'm out in my Somerset garden: it's part of the sonic landscape of the season in these rural parts, blending into the background, rather than forcing its way into my consciousness.

If a male does manage to arouse a female's attention, he will display again where he plans to build a nest, fanning his tail and singing until she finally shows an interest, and lands on a nearby beam or wall. Immediately he approaches her, uttering affectionate sounds to tempt her to pair up with him.

As with other birds, the male appears to have the upper hand: he, after all, is using all his efforts to produce such stirring and persistent aural and visual displays. But it is the female who actually does the choosing and so, despite appearances, is the dominant partner. For the male, there is no guarantee of success: it can take as long as a month to pair up – and some never do.

Exactly how female birds choose their mate has long occupied the minds of scientists. For some species, it is the quality, variety and persistence of the male's song; for others, it might be the brightness of his plumage. Both indicate that the male is in good health, and so will not only be a good father to any offspring but will also pass on the same high-quality genes.

While the male swallow's song and plumage (especially the white tail spots) are influential, the deciding factor appears to be the length of his tail. All other things being equal, males with longer, straighter and more symmetrical tails are usually more successful in getting a mate. They also tend to win over a female more quickly. The reason is simple: the extra energy needed to grow a long tail shows the bird is fit and healthy; a sick bird (especially one infested with parasites like feather lice) will not be able to spare the energy needed to produce a longer tail.

By choosing a long-tailed male, the female is increasing her chances of success, in both the short and long term. He will make a better parent, being able to bring back more high-quality food to their chicks; he will also pass this characteristic on to their male offspring, making them too more likely to be more successful in winning a mate and raising a family, and eventually a whole dynasty of swallows down the generations.

Yet this raises a tricky question. If longer-tailed males are indeed more successful, surely over time, through a self-reinforcing process known as runaway sexual selection, we would have seen male swallows' tails become ludicrously long – like a peacock's?

That this has not happened might suggest there comes a point when a long tail becomes more of a disadvantage than an asset. Too short a tail, and a male may not attract a mate at all; too long, and he risks being unable to catch enough food, as his manoeuvrability – the ability to twist and turn to grab fast-flying insects in mid-air – might be compromised. A tail too long and heavy would also make him more likely to be caught by an aerial predator such as a hobby or sparrowhawk.

Once the female has chosen both a male and a suitable nest site, it is time to get down to the serious business of breeding. Swallows mate briefly and often, the male hovering over his partner with his tail feathers fanned out (presumably to give him more stability), while she adopts a submissive position, back held horizontal and wings lowered, allowing him to copulate with her.

Now the hard work really begins. Over the next few weeks the male and female will build a new nest (or repair an existing one), where the female can lay her eggs. Once these have hatched, the male and female will raise their young before, when autumn comes, heading south to Africa once again.

This race to reproduce is a long and arduous process, consuming almost the swallows' entire stay with us, and ultimately determining whether or not that long journey north has been worthwhile.

Birds that live all or part of their lives alongside human beings, like the house sparrow, barn owl and, of course, the barn swallow, are known as 'commensal'. In the case of swallows, they depend on us not just for places to breed, but also for food. Their preference for nesting in farmyards is closely tied to the presence of horses and cattle, whose plentiful dung acts as a kind of food magnet, attracting an abundance of flying insects. Indeed, it could be argued that the creatures with which they are truly commensal are not us, but our livestock.

This also suggests that in the not-too-distant past, when swallows nested in natural sites in caves or on rock faces, they would, as Collingwood Ingram pointed out, have been neither as common nor as widespread as they are today: 'Because of the relative scarcity of such places in the remote past the swallow must necessarily have been a comparatively rare bird, or, at any rate, a very locally distributed species.'

Only around 10,000 years ago, when the shift from hunter-gathering towards agriculture allowed formerly nomadic peoples to stay in one place and create permanent settlements, would swallows have switched to breeding in and around human habitations. At first, they would probably have nested in people's homes, all-purpose structures housing animals as well as human families, which provided a relatively warm and safe place where the swallows could raise their young. Today, they nest almost exclusively in our farm buildings; especially, and appropriately given the official name, in barns – in what Ingram described as 'a unilateral symbiotic relationship'.

Almost all nest sites – at least in the UK and at other more northerly latitudes – are effectively 'indoors'. Elsewhere, in the warmer, more southerly parts of their range, swallows do nest in outdoor locations: beneath bridges or, like house martins, under the eaves of houses. In North America, they continued to use natural nest sites for rather longer than in Europe, because the mass colonisation by human beings – and their settlements – occurred far later on that side of the Atlantic.

The best sites will tend to attract more breeding pairs, and while the swallow only forms loose colonies, pairs do often nest fairly close together. That has drawbacks as well as advantages: the male needs to guard against his chosen female mating with his rivals; on the other hand, he may also have the opportunity for some extra-curricular activity, while more pairs of eyes make the birds safer from predators.

In barns, the most popular nest sites are on top of wooden beams, whose usually broad, horizontal surface is ideal for creating a stable nest that will not easily fall off. Living in a working village, John Clare was very familiar with this:

> Swallows … build on the beam of a shed … which supports the roof by running from end to end of the building – they [the nests] have a very odd appearance & are placed in the same manner as a saucer on a mantel piece or a basin on a shelf & look exactly as if put there.

True to their now obsolete name of chimney swallow, they also, as Gilbert White noted, used to build their nests in working chimneys: '[Swallows] love to haunt those stacks where there is a constant fire, no doubt for the sake of warmth. Not that it can subsist in the immediate shaft where there is a fire; but prefers one adjoining to that of the kitchen.' Even so, White observed that 'the swallow, though called the chimney-swallow, by no means builds altogether in chimneys, but often within barns and outhouses against the rafters, and so she did in Virgil's time: "The tottering swallow hangs her nest from the rafters."'

The Victorian writer and ornithologist William Yarrell also described the very varied choices made by swallows when it came to choosing a nest site:

> In the north of England these birds frequently build in the unused shafts of mines, or in old wells; sometimes under the roof of a barn or open shed, between the rafters and the thatch or tiles ... Turrets intended for bells are frequently resorted to, and unused rooms of passages in outhouses, to which access can be gained by the round hole so frequently cut in the doors to such buildings, and within which the birds take advantage of any projecting leg, or end of a beam, that will serve as a buttress to support the nest.

Swallows will build their nests almost anywhere else inside these often cluttered and disorderly buildings, as Angela Turner notes, so long as there is something for them to fix the structure onto:

> Objects attached to a wall or hanging from a roof or beam also provide support: these have included wasps' nests, old birds' nests ... picture frames, hats, pieces of cloth, lampshades and brackets, electric wires, masonry bolts, nails, chains, gear wheels and pulleys, a sunflower seed-head nailed to a beam and the corpse of an owl!

The last of these requires a fuller explanation, provided by Gilbert White:

> Stranger still, another bird ... built its nest on the wings and body of an owl that happened by accident to hang dead and dry from the rafter of a barn. This owl, with the nest on its wings, and the eggs in the nest, was bought as a curiosity worthy of the most elegant private museum in Great Britain [that of Sir Ashton Lever]. The owner, struck with the oddity of the sight, furnished the bringer with a large shell, or conch, desiring him to fix it just where the owl hung: the person did as he was ordered, and the following year a pair, probably the same pair, built their nest in the conch, and laid their eggs.

David Bannerman cited several unusual nest sites, including – as observed by the pioneering duo of Scottish women ornithologists Evelyn Baxter and Leonora Rintoul – a pair that nested in an old fishing boat at Aberlady in Scotland, the nest being only two feet above the water level at high tide.

One contributor to the RSPB's online forum posted an observation of a nest underneath the pontoons at Chichester Marina in West Sussex, also just a couple of feet above the high-water mark. They were concerned that when the young fledged, they would drop into the water and drown, though I suspect the chicks, which can usually fly pretty well when they leave the nest, were able to avoid this fate.

The swallow's nest is a shallow half-cup made from individual pellets of mud (or dung), mixed with pieces of grass, straw and horsehair, all assembled together to create a rather untidy-looking structure. Nevertheless, this is strong enough to hold not only a clutch of eggs but, later on, also a brood of large, active, almost-fledged chicks.

The female does the lion's share of the nest-building, but the males do lend a hand: each bird making as many as thirty trips every hour – perhaps a thousand in all – to bring back mud and other nesting material. They usually do the bulk of the construction work in the morning, so that the mud has dried by the end of the day.

In recent years, both swallows and house martins have found nest-building trickier, because of a lack of mud around where they breed. Building sites are much tidier than they used to be,

while more than half a million farm ponds across Britain have been drained and filled in since the Second World War. Recent spring droughts have also led to shortages of mud, as in 2011, when the driest April on record across much of the UK meant that many swallows and house martins were unable to build new nests.

To get around this problem, and save time and effort, swallows will often reuse old nests, doing basic repairs to make them more habitable. One study showed that almost half the swallows in a colony reused the same nest as the year before. However, the benefits of doing so may be offset by the presence of parasites that have overwintered in the old nests.

The same nest can be used for several years in a row: indeed, the longest recorded occupancy of a single nest, in Germany, was an astonishing forty-eight years, by successive generations of the same family.

Sometimes, however, the resident birds have beaten the swallows to it. The Victorian nature writer Edward Stanley, Bishop of Norwich, recalled one example of unexpected squatters, which met with a brutal response:

> A pair of Swallows … on arriving, found their old nest already occupied by a Sparrow, which kept the poor birds at a distance, by pecking at them with its strong beak, whenever they attempted to dislodge it … They, at last, hit upon a plan which effectually prevented the intruder from reaping the reward of his roguery. One morning they appeared with a few more Swallows – their mouths distended with a supply of tempered clay – and by joint labour,

> in a short time, actually plastered up the entrance-hole, thus punishing the Sparrow with imprisonment, and death by starvation.

Stanley, however, was notorious for reproducing false and unverified stories – including 'Eagles carrying off children' – so we might take all this with a large pinch of salt.

Once the main structure of the nest is complete – after anything between five and twelve days, depending on the weather – the birds line it with grass and a layer of hair and feathers, ready for the female to lay her first clutch of eggs. Most nests are at least two metres above the ground, and often several metres higher, to keep the eggs and chicks (and incubating female) safe from predators, especially farmyard cats.

Another key requirement of a suitable nest site is that the adult birds are able to fly in and out of the barn or outbuilding with ease. Often these buildings will in any case be open to the elements, with a large door or entrance for livestock and farmworkers to get in and out. In closed structures, the swallows can usually find a small hole or gap – as little as five or six centimetres in diameter – through which they can fly at terrifying speed.

The ability of nesting swallows to gain access to buildings can cause problems for the owners. When the author and broadcaster Malcolm Welshman replaced his ramshackle old workshop with a spanking-new one, he hoped that his regular pair of swallows would nest instead in his car port, just a short distance away.

But he had forgotten that swallows have an extraordinary loyalty not just to the general location of where they have previously nested, but to exactly the same site. 'I should have sensed their annoyance,' he explains,

> in the angry *'tswit-tswit'* that echoed through the air as they circled the new car port, chose to ignore it and bombed the new workshop door, now firmly closed and strictly out of bounds. But I had to open the door at some point, and as soon as I did, the swallows seemed to appear from nowhere and with a gleeful whoosh over my head, zoom in before me.

That night, he locked the workshop door, assuming that the birds would go elsewhere. Yet the following morning, to his surprise, they were back inside, busily building a nest. They had found a tiny gap around the eaves, and somehow squeezed through. In the end, he decided that their tenacity deserved reward, and allowed them to nest in his pristine new workshop.

Like anyone else who has had the privilege of swallows nesting close by, Welshman enjoyed witnessing them raise their young:

> At the end of the summer, our own swallow family lined up on the telegraph wire in readiness for leaving: a row of chattering, lively youngsters and their parents. Begrudgingly, I had to admire those parents. They'd fought and won their battle against me. Now they had another battle to overcome in their arduous migration south. 'Safe journey,' I whispered as the little flock finally rose and, with a last chorus of calls, circled the workshop and disappeared over the fields …

The following spring, the swallows did return – to be welcomed 'with open arms – and an open workshop door!'

That several species of swallows and martins nest in, on, or near our homes has inevitably led to them being regarded as good omens: not only do we look after them, but they in turn protect us. The Chinese believe that swallows nesting in a dwelling bring good luck to the human residents, while traditional Chinese poetry often depicted a woman's voice as the twittering of a swallow.

Likewise, destroying a swallow's nest is often seen as bringing misfortune. One widespread belief holds that if the eggs are taken from a nest, the milk produced by the cows in that barn will be stained with blood. Worse still, such mindless vandalism might lead to the death of the perpetrator or a member of their family. Killing a swallow is also taboo, presumably because it is seen as a breach of trust, given that these birds choose to nest so close to where we live.

But swallows are not always seen as a good omen. In Shakespeare's *Antony and Cleopatra*, Scarus, one of Antony's commanders, has warned him not to fight his rival Octavian by sea. He is worried that, in a departure from their usual habits, swallows have built their nests in the sails of Cleopatra's ships. His misgivings prove prescient: Cleopatra betrays Antony and flees with her forces, and the battle is lost.

It's not the only example of how a bird that is generally regarded as a positive symbol of renewal can sometimes turn into a portent of doom – especially if it deviates from its typical behaviour. The same is true of the robin, which is welcome on our doorstep but, as soon as it crosses the threshold and into our home, is viewed as an ill omen, foreshadowing the death of one of the occupants.

Even today, not everyone welcomes the return of the swallow. In March 2019, the *Guardian*'s Kate Blincoe revealed that the supermarket giant Tesco had destroyed a nesting site for swallows at its Norwich superstore, following customer complaints about bird droppings on their shopping trolleys. Having used power hoses to wash away the nests, Tesco then fitted netting to ensure that the swallows, which were about to return, would not be able to enter.

In this case, social media triumphed: Kate's Twitter post was retweeted many thousands of times, kickstarting a proposed customer boycott of Tesco. A magazine in South Africa even picked up on the story – as Kate pointed out, 'they were their swallows, too.' Tesco soon backed down and removed the nets. This was no isolated incident: elsewhere in Norfolk a sand martin colony along coastal cliffs was also netted before the birds had returned.

Hostility to swallows is not, it seems, confined to this side of the Atlantic. In 2017, US farming website Agweek ran an article headlined 'Always in season: Barn swallows are a love–hate species.' The author, Mike Jacobs, began by acknowledging the attractiveness of swallows, but soon changed his tune: 'The barn swallow is a pretty little bird. It can also be an aggressive little pest.' He went on to accuse nesting swallows of being 'trespassers', who 'don't respect our rules of property', and advised his readers to knock down any nests – which would be illegal in the UK but appears to be acceptable in the US. Finally, he complained about the birds' aggressive nature – before grudgingly admitting that he admires their beauty and sociability.

One US pest control firm, Bird B Gone, sells various products designed to 'get rid of swallows', including netting and visual and sound deterrents, such as the 'Bird Chase Super Sonic', which plays the birds' distress calls. It also advises that you 'discourage the feeding of swallows by children or employees', which shows a lamentable lack of understanding of a swallow's diet and feeding methods.

What all these stories reveal is humanity's growing disconnection from the natural world; a disconnection that ultimately threatens us all. To counter this hostility, Kate Blincoe ended with a plea for tolerance and understanding:

> We need to celebrate our contact with wildlife, which increasingly has to live alongside us in urban and suburban places. Here's hoping for webcams, not nets; for schoolchildren visiting, instead of pest control; and beautiful vertical gardens and swallow nest cups to replace the blank sterility of a pressure-washed wall.

As with other songbirds, the number of eggs laid by swallows does vary: the usual clutch is four or five, but it can be as few as two, or as many as eight. Birds that nest farther north tend to have larger clutches – which makes sense, given that swallows originally travelled here to take advantage of the longer hours of daylight and greater abundance of insects. However, second clutches laid by these birds are usually smaller as by that stage in the year, with autumn approaching, the hours of daylight are rapidly shortening.

These northerly populations also begin nesting far later than more southerly ones: in May or June in Scandinavia, compared with February or March around the Mediterranean, and April or May in the UK. Even here, timing will vary: those in northern England and Scotland lay, on average, a week or so later than those in the south. As always, this depends on the weather in any particular year, and at each specific location.

For swallows, as with other birds, there is also a trade-off in terms of the number of eggs and chicks: between what they could theoretically have in ideal conditions, and the number they can actually raise. The deciding factor is the unpredictability of the weather, which in turn affects the supply of flying insects.

The eggs themselves are off-white in colour, with variable reddish-brown spotting, the spots being bigger and more densely packed at the larger, rounded end. Like almost all songbirds, the female lays a single egg each day – usually early in the morning – and does not start to incubate until the whole clutch is complete, thus ensuring that the chicks will all hatch out at roughly the same time.

She incubates her eggs for between 13 and 16 days, with no help at all from her mate (in Britain and Europe at least; in North America male swallows do take some part in the incubation duties). But once the chicks hatch – naked, blind and helpless, and each weighing less than two grams – the male redeems himself by taking a full part in their feeding. Indeed, of all the world's 5,000 or more songbirds, male swallows and martins make the greatest contribution to caring for the young.

They certainly have their work cut out. A single brood of baby swallows needs to eat an astonishing 150,000 individual insects before the chicks fledge, some 18 to 23 days after hatching: that's between 7,000 and 8,000 insects a day, or as many as 500 every single hour. Most are caught during the middle part of the day, when air temperatures – and therefore insect activity – tend to be higher. No wonder Gilbert White praised swallows as benevolent pest controllers:

> A most inoffensive, harmless, entertaining, social, and useful tribe of birds: they touch no fruit in our gardens ... amuse us with their migrations, songs, and marvellous agility; and clear our outlets from the annoyance of gnats and other troublesome insects.

Swallow parents don't just find food for their chicks: they also have to feed themselves to keep their energy levels up, and perform other duties, such as removing fragments of eggshells and the chicks' faecal sacs (memorably described by Bill Oddie as 'shrink-wrapped poo') from the nest, to maintain good hygiene. This goes on for the first few days after the chicks have hatched; after that, the chicks simply deposit their droppings over the rim of the nest, producing tell-tale spatters of white on anything below.

When catching food for their young, the busy parents swoop around and grab their tiny prey out of the ether, before bringing a bolus (concentrated ball) of a hundred or more insects back to the nest, sometimes at the rate of one visit every minute. However, their strike-rate is highly dependent on the prevailing weather conditions: swallows can usually find many more insects on a sunny day, but in poor weather the adults may be away from the nest for far longer.

As soon as either parent arrives back at the nest, the baby swallows spring into action, as if prompted by an electric shock: rising up and opening their beaks as wide as possible, to reveal a bright, custard-yellow gape. This, along with their loud begging calls, stimulates the parents to provide as much food as they can. It often appears that the larger and more dominant young get more food, but adult swallows are very good at making sure they share out their booty between every member of their brood, to ensure they all grow at the same rate and fledge successfully.

Feeding the young can sometimes be a tricky business. In the first volume of the journal *British Birds*, published in 1907–8, one observer wrote an account – complete with accompanying photograph – of a nest built on one side of a gas shade (shaped 'like an inverted soup plate') in which a pair of swallows had successfully reared two broods during the summer of 1907:

> When the young … clamoured for food, it was an extraordinary sight to see how the shade would swing about, and, owing to the weight being on one side, would often reach an angle at which it became quite dangerous for the occupants of the nest … In spite of the slippery nature of the glass on which the nest was built, it appears to be very securely fixed.

The manner in which an adult swallow flies into the nest at great speed and manages to brake and land, all without dropping its precious cargo of food, is easy to miss: the bird is so fast it goes by in a blur. Such aerobatic skills at the nest were first appreciated by the pioneering bird photographer Eric Hosking. In his autobiography *An Eye for a Bird*, he described the difficulties of photographing swallows in the spring of 1948 at Staverton Park, near Woodbridge in Suffolk.

> I was attempting to photograph birds in flight, but all I achieved was a discovery of another snag about high-speed photography

> ... A small bird flying about four feet in front of the camera at 15 mph is in the field of view of the lens for only about one-tenth of a second. And in that fraction of time I had to decide if the bird was in focus, and press the shutter release at the precise moment when it would be in the centre of the plate. Coordination of eye, brain and finger were just not swift enough. Consequently, my first results were a fine crop of disappearing tails, half heads, empty bushes and clear skies.

Hosking sought the advice of his friend and fellow bird photographer, an electrical engineer named Dr Philip Henry. Henry astutely realised that 'the birds must photograph themselves', and found a clever solution to the problem. He developed a photo-electric beam which, when broken by the arrival of the swallow, simultaneously triggered the flash and the shutter, taking a photograph at exactly the right moment. as Hosking relates:

> Swallows arrived with mouths crammed with flies caught on the wing and, as they slowed down immediately before alighting, were caught in some delightful poses. These photographs revealed, probably for the first time, the swallow's perfect streamlining. At first sight they appeared to have no legs or feet, until it was realised that these were tucked under the stomach feathers, thus reducing wind resistance.

Later, Hosking used the same device to take what is his best-known picture: a barn owl returning to its nest, wings outstretched, carrying a vole in its beak – an image that has been reproduced in more than a hundred different countries.

With so much food being brought back to the nest, no wonder swallow chicks grow at such a rapid rate. After just 12–15 days, their weight has increased as much as twelve-fold, reaching between 22 and 25 grams (almost one ounce) – noticeably heavier than their parents. This extra weight, in the form of fat reserves, enables them to survive during periods of bad weather, when insect food may be temporarily scarce. During the final few days in the nest, their weight begins to fall until, when they actually fledge, they weigh between 17 and 20 grams, about the same as a typical adult bird.

There is another good reason for this growth spurt, and also the relatively long time baby swallows spend in the nest compared with other songbirds (usually three weeks or so, compared with as little as ten to eleven days for some other species). They have to be able to fly – and hunt for aerial food – as soon as they leave, so they need their flight muscles to be fully developed.

By the end of their time in the nest, a brood of swallows is very obvious: the youngsters are now standing up when they sense their parents returning, making an even greater cacophony. This is a risky thing to do, as it can alert predators such as cats and sparrowhawks to their presence. Fortunately, both parents will defend their offspring with great energy and determination, uttering loud alarm calls to alert their chicks and neighbours, and often chasing the attacker away; the young, meanwhile, will keep quiet and crouch as low as they can inside their nest.

Roughly three weeks after they hatched, the chicks could hardly look more different. Their downy fluff has now been replaced with a full set of wing, tail and body feathers and, while they are noticeably drabber in colour than their parents, the blue back and tail, pale buffish underparts and russet throat are all clearly visible. And as they jostle one another in an increasingly tight fight for space, each perhaps senses that the moment of truth – when they have to leave the safety and security of the nest for the first time – is getting ever closer.

We cannot begin to imagine the consciousness of a swallow chick as it prepares to take its first leap into the unknown.

Remember, until now all it and its siblings have ever experienced is their immediate surroundings: the nest itself, and the limited view when, as they grow older and stronger, they are finally able to peer over the rim.

Unlike swifts, which live in near darkness under the roof tiles or eaves of buildings, and whose view of the world, as they stare down a tunnel to a circle of light in the distance, is even more limited, swallows can at least see around the barn in which their parents have chosen to nest. There may be bales of straw, a tractor or other farm machinery, perhaps a cow or two. But the great outdoors remains unseen.

Like all baby birds, a swallow leaves the nest when instinct tells it to; when its flight feathers have developed so that it can stay airborne for long enough to escape the confines of the barn and take to the air.

When this does finally happen – after plenty of hesitations and false starts – it looks surprisingly easy. First, the young swallow flexes its legs to raise it up and over the rim of the nest; then it spreads its wings, leans forward like a high diver on a springboard, and gravity does the rest. One moment the youngster is there, the next gone, swooping up into the late spring skies on those broad, pointed wings, and experiencing this bright new world for the very first time.

Once this pioneering chick has left the nest, its siblings follow; though it can take another 24 hours or more – sometimes as long as three or four days – before they have all taken the plunge.

When swallows nested more commonly in chimneys, leaving the nest could prove hazardous, especially if the youngster fell out before it had fully fledged. 'Today', noted John Clare (it was 21 July),

> a young Swallow fell down the chimney where it lay chirping a good while – at last the parent birds ventured down and fed it – & a short time after 2 other swallows joined them & by some means or other got the young one up to its nest in the chimney top.

Once the young are safely out, they usually remain close to the nest site, perching in a tree or on a telegraph wire, where they feel safe. The male and female continue to bring them food – and feed them beak-to-beak as before – for about a week afterwards, and sometimes longer; until the youngsters have learned to catch prey for themselves.

So far, I might have given a somewhat misleading impression of swallow family life. It all appears rather delightful and harmonious: the devotion of the parents to their chicks when in and out of the nest; the tranquillity of the scene, enhanced by the gentle twittering of the adult birds as they hawk for insects under a sunny sky …

Nothing could be further from the truth. Although swallows do generally form monogamous pairs, which sometimes stay together from one year to the next, when it comes to breeding behaviour they, like many small birds, engage in all kinds of shenanigans. Although they appear to be faithful to their mate, both males and females are always on the look-out for opportunities to maximise their chances of breeding success.

A clear imbalance between the sexes here is expressed in very different strategies and tactics. If a male manages to successfully mate with a second female, he need take no further part in raising his extra-marital offspring, as her own mate will unwittingly do all the hard work. So, for him, being promiscuous has little or no costs, and many benefits.

The female, by contrast, can only lay a single clutch of eggs, and raise one brood of chicks, at a time. But she can still hedge her bets by mating with other males. By laying a clutch fertilised by two or three different fathers, she is not putting all her eggs, as it were, in one basket.

Angela Turner points out that if a female arrives back late from her winter quarters, and all the 'best' males are already taken, she can still improve her chances of raising stronger young by sneaking off to copulate with other, healthier males. However, her mate will of course be alert to this, and guard her assiduously, especially during the period before the full clutch is laid, when the females are most fertile.

When several pairs of swallows are nesting together in a loose colony, as they often do, males may be seen chasing other females (to mate with them) and other males (to stop them mating with their partner). Scientists estimate that between a third and half of all nests contain at least one chick which does not belong to the incumbent male. It's a wise swallow, it seems, that knows its own father.

If this is starting to sound a bit like the avian equivalent of a certain kind of daytime TV show, that's because it is. And it gets worse. Male swallows are known to occasionally practise infanticide: arriving at a nest and ejecting the incumbent chicks by picking them up and throwing them out to die. This darker side of swallow behaviour reached a wider audience when, in June 2008, millions of people tuning in to the popular wildlife soap-opera *Springwatch* were confronted with a shocking incident.

From their base at Pensthorpe Natural Park in Norfolk, presenters Bill Oddie and Kate Humble were relaying footage of a swallow nestbox, with a very fidgety pair of adult birds noisily squabbling with one another. Soon afterwards, the four chicks hatched, and the female began to feed them. Then the male arrived but, instead of bringing food, he uttered a loud and urgent note, chased his mate away, plucked a tiny, naked chick out of the nest, and dropped it to the ground. The female subsequently returned, and immediately the male came back and mated with her. Over the next few hours, he took away the remaining three chicks, which, like their unfortunate and helpless sibling, soon died. The male then built an entirely new nest in the same building, on top of one of the remote camera brackets, a few metres away from the original nestbox; and he and the female went on to successfully rear a new brood of young.

It may have been surprising to see, but it is not uncommon: the British Trust for Ornithology estimates that infanticide occurs in between 4 and 5 per cent of swallow nests. It usually happens when for some reason the original male has disappeared or died, at a time when his brood of chicks are still very young and vulnerable; but male interlopers sometimes kill chicks when both the original male and female are still feeding them.

In all his years of watching birds Bill Oddie had never

witnessed this shocking behaviour before. He correctly compared this swallow to male lions that, when they infiltrate a pride and drive away the existing males, often kill all the cubs, to ensure that the previous male's genetic lineage comes to an end. It may be grisly, but it does make evolutionary sense.

Because swallows spend so much of their lives living alongside people, they tend to attract greater attention than many other small birds. This was especially the case during the First World War, when swallows continued to build their nests and raise their broods amid the noise and horror of the Western Front.

Fortunately, a number of keen birdwatchers were stationed there, who carefully documented the behaviour of the birds that made their homes alongside them. Nesting swallows soon took on a significance beyond the purely avian: for the more reflective and thoughtful servicemen, they came to stand for a kind of normality.

Later in life Charles Raven would become a distinguished theologian, author and Master of Christ's College, Cambridge. But during the First World War he was a young chaplain on the front line in France, witnessing such terrible scenes that, after the conflict ended, he became a lifelong pacifist. In his biography, *In Praise of Birds*, published less than a decade later, Raven wrote of how the discovery of a pair of swallows nesting in the troops' temporary HQ was a morale boost for his men: 'These birds were angels in disguise. It is a truism that one touch of nature makes the whole world kin; those blessed birds brought instant relief to the nerves and tempers of the mess ... we all regarded the pair with devoted affection.' The swallows were soon adopted as the battalion's unofficial mascots, their survival clearly linked – in the minds of Raven's men at least – with their own fragile mortality.

Another soldier who made careful observations of swallows on the Western Front was Hugh Gladstone – or, to give him his full name, Sir Hugh Steuart Gladstone of Capenoch. A scion of the Scottish landed gentry, educated at Eton and Cambridge, Gladstone had served with distinction in the Boer War. When the First World War broke out, he was in his late thirties, and could have easily been excused from the fighting and stayed at home. But he chose to rejoin his regiment, the King's Own Scottish Borderers, as a captain, and fight on the front line with men less than half his age.

In 1919, just a year after the Armistice, Gladstone published a slim volume of observations entitled *Birds and the War*. Not surprisingly, perhaps, swallows feature frequently in its pages:

> 'No-man's-land' proved an attractive place, in spite of the noise and all the dangers of artillery fire, for thousands of birds to nest and rear their young. Nor must it be forgotten that the abnormal quantity of insects doubtless formed an attraction to insectivorous birds, and this was particularly noticed as regards Swallows, Martins, and Swifts.

That 'abnormal quantity of insects', one has to assume, would have been the consequence of the equally abnormal quantity of lifeless, decomposing flesh.

Like Charles Raven, Hugh Gladstone noted how the swallows soon adapted to the unfamiliar landscape. They were apparently unfazed by all the frenetic activity, often nesting in the Nissen huts put up as temporary accommodation, and in field dressing stations, where they were 'unusually tame and confiding'. They also coped well with the noise and disturbance:

'Another pair built their nest in an "Armstrong hut" only 60 or 70

yards from a battery of 6-inch howitzers, which fired at intervals of about three minutes or less throughout the day, and on special occasions all night long.'

The birds did, Gladstone thought, react to the sound of anti-aircraft guns, appearing to fly 'more spasmodically than usual'. It was likely, he suggested, that they were disturbed not by the sound of the explosions, but by the vibrations in the air as the shock of the blast dispersed.

Generally, though, swallows appeared remarkably unaffected by the chaos going on all around them. Gladstone observed that one pair did not desert their chicks when a shell took off the roof of the building in which they were nesting, and that 'before the day was over they used the shell-holes as a convenient entrance through which to pass backwards and forwards with food for their young.' Another pair tried to construct a nest under the hood of a lorry, continuing to add nesting material every time the vehicle returned to base; however, after several days they gave up.

Later that year, Gladstone noticed a flock of swallows 'quietly perched on war-telegraph wires, before migrating, as though the turmoil of battle were a thousand miles away'. In a poignant symbol of the destruction of war, he also saw, in the spring of 1918, swallows and house martins circling around the cathedral and ancient church tower at Ypres ('Wipers' to the English Tommies):

'These were daily shelled and hit; none the less, nest building went on with patience and perseverance, the necessary mud being obtained from very old shell-holes and the canal banks.'

Another meticulous observer was Collingwood Ingram,

whose wartime diaries *Wings over the Western Front* – which include numerous observations of birds – were not published until 2014, a full century after the start of the conflict.

Collingwood 'Cherry' Ingram gained fame in later life as a plant collector and horticulturalist, earning his unusual nickname from his expertise on Japanese flowering cherries. So important was he in promoting these special trees that his 2019 biography is subtitled 'The Englishman who saved Japan's Blossoms'. But along with his passion for flowering cherries, Ingram was a lifelong ornithologist, and member of the British Ornithologists' Union for a record 81 years until his death in 1981, at the age of 100. I have already quoted from his delightful book *The Migration of the Swallow* (published in 1974, in his ninety-fourth year). But now let us concentrate on those diaries, compiled during the two years he served in France with the Royal Flying Corps, from December 1916 until his final entry, on New Year's Eve 1918.

Many entries are simple observations – capturing the immediacy and excitement of newly arrived birds in spring, such as this entry from 10 April 1918, in which Ingram welcomed back the first swallow with undisguised delight:

> Today a swallow flew over, twittering joyously in the bright sunlight as he glanced across the turquoise sky. The first swallow always imparts a thrill of pleasure and is a red-letter day for all lovers of birds.

Perhaps because the Second World War was fought on a less human, more industrial scale, there is far less bird-based literature from that conflict. One notable exception is *Wing to Wing: Bird Watching Adventures at Home and Abroad with the RAF*, by E. H. Ware. Set mostly in North Africa, it features this detailed account of the problems facing breeding swallows in the intense heat of Bordj Menaïel, in the western Kabylie region of Algeria:

> Two pairs of Swallows nested in our billet, and were a great source of interest to the men. The schoolroom being a modern flat-roofed, concrete-built structure, there were no beams for them to build on in the ordinary way. The nearest approach … were the long, square concrete 'joists' directly supporting the roof. One pair of birds attempted to build on the side of one of these, House Martin fashion, and after much plastering succeeded in half-building a nest. They then evidently decided that it was too much like work and gave up the attempt. The other pair were wiser, and chose the top of an electric lampshade, close under the ceiling. They completed their nest, laid, and sat for a long period, but the eggs proved infertile, possibly owing to the great heat, the room itself getting very warm during the summer and the lamp being in use most of the day.

The most detailed observations of swallows during the Second World War came from two prisoner-of-war camps in Bavaria, Laufen and Eichstätt, to which migrant birds would return from Africa each spring – starkly pointing up the prisoners' own lack of freedom.

By coincidence, several keen ornithologists and birdwatchers found themselves incarcerated there. Led by the scholar and poet John Buxton, they made a number of important studies of the camps' birdlife, including what became the monograph on the redstart Buxton published after the war, in Collins' New Naturalist series. Their extraordinary, and life-affirming, story is told by Derek Niemann in *Birds in a Cage*.

One prisoner, Richard (Dick) Purchon, described by Niemann as Buxton's 'aide-de-birdwatching', chose to study another spring and summer visitor to Eichstätt, the swallow. With military precision, Purchon organised a team to watch and record the behaviour of these birds around the clock. The 15–20 pairs of swallows nesting in the camp buildings proved an easy subject for study, as Purchon pointed out in a paper published in 1947: 'During the frequent air-raid alarms prisoners-of-war were without exception confined to barracks and under these conditions swallows were especially suitable subjects for observation.'

One male swallow's inspection of potential places to nest was described by a fellow inmate, Peter Conder: 'He comes in, hovers round with slow wing beats, tail hanging down, settles on the pipes containing electric wires, now glides, settles on the lampshade … It hovers very slowly, going up and down stairs and flies out by the upper landing window.'

Keen to learn more about the swallows' migratory habits, Purchon and Buxton decided to ring the chicks, using homemade metal bands placed around their legs. Incredibly, these survived the journey to Africa and back, and the following spring, no fewer than seven of the ringed swallows returned to the very same place they had been born.

It was not all plain sailing for the swallows. One nest fell down from the rafters, splitting open on the floor to reveal a brood of chicks. Despite their frantic cheeping, their parents kept returning to the original nest site, seemingly unable to understand that their offspring were lying helpless just a few feet below. The ingenious Buxton made a substitute nest out of an empty tin of condensed milk, lined with cotton wool, into which he put the chicks, replacing the new 'nest' back on the same rafter. This makeshift structure worked: the adults began feeding the chicks which, none the worse for their adventure, eventually managed to fledge.

After the war, Peter Conder gave up a career in advertising and joined the RSPB, ultimately becoming the Society's chief executive. Dick Purchon, who before the war had been a high-flying young zoologist, returned to academia, where he enjoyed a distinguished career as a marine biologist before his death in 1992. And John Buxton went back to New College, Oxford, where he taught English Literature to generations of students.

Purchon's swallows, along with Buxton's redstarts, took on an importance way beyond subjects for ornithological studies, or a means of staving off boredom. As Buxton noted, all the migrant birds that visited them each spring and summer became a living embodiment of the freedom for which these men had been fighting, now so threatened; as well as a simple focus of hope:

> But one of the chief joys of watching them in prison was that they inhabited another world than I ... They lived wholly and enviably to themselves, unconcerned in our fatuous politics

> ... They lived only in the moment, without foresight and with memory only of things of immediate practical concern to them ... memory also, perhaps, of the way back, when their one necessary purpose was done, to the hot sun of Africa.

Implicit in this powerful passage about the redstart and what a modern audience might call its instinctive 'mindfulness' is how migrant birds like it and the swallow have, since ancient times, been treasured, above all, for that sense of 'elsewhere' – currently inaccessible to us humans, maybe, but embodied in them as an eternal possibility and aspiration. But for now, and during the coming season, they are right here with us: a constant presence throughout our all-too-brief northern summer.

3

SUMMER

For one swallow does not make a summer, nor does one day; and so too one day, or a short time, does not make a man blessed and happy.

Aristotle, *Nicomachean Ethics*
(fourth century BC)

On what would turn out to be yet another record-breaking hottest-ever day in the UK, I took our fox-red labrador Rosie for an early-morning walk along the lane behind our home. It was already very warm and in such fine, settled weather the swallows, as I expected, were out in force.

They were not the only ones. The sun-bleached meadow brown butterflies, fluttering weakly across the lane from hedgerow to hedgerow, looked as if they could barely find the energy to flap their wings. A more energetic greenfinch launched himself into a late-season song-flight before parachuting down to a hawthorn twig; while a distant bullfinch uttered his soft, piping call.

The swallows swooped and dived around me, picking up invisible insects just above the newly mown hayfield. As I looked across at our parish church tower it struck me, not for the first time, how some birds, like the house sparrow and the barn owl, spend their whole lives in our little village, while the swallows truly are global wanderers.

They would still be with us for a couple of months more. Yet as my wife Suzanne and I were having breakfast on our balcony that morning, four swifts had swept overhead, heading purposefully south on their journey to Africa. These were the first – and probably last – swifts I had seen over the garden that year, for in Somerset they are mainly a bird of towns and cities, and only occasionally fly over our village.

The newspapers were full of photos of people enjoying the hot weather, together with the usual variations on the 'Phew – What a Scorcher!' headline. But this summer, for the first time, I detected a different tone to the coverage of the heatwave: more concerned than celebratory, reflecting a new awareness of the perils of the global climate crisis. It made me wonder about the fate of our migrating swallows in the future, when many of their stopover points and wintering areas are likely to become increasingly hot, dry and hostile.

The British summer – at least in our lowland countryside – is closely associated with the presence of swallows. As Gilbert White observed, from dawn to dusk, the parent birds can be observed collecting food for their young:

> All summer long is the swallow a most instructive pattern of unwearied industry and affection; for, from morning to night, while there is a family to be supported, she spends the whole day in skimming close to the ground, and exerting the most sudden turns and quick evolutions. When a fly is taken a smart snap from her bill is heard, resembling the noise at the shutting of a watch-case; but the motion of the mandibles is too quick for the eye.

Swallows, White noted, will often follow horses, which, like grazing cattle, attract their fair share of flies: 'Horsemen on wide downs are often closely attended by a little party of swallows for miles together, which plays before and behind them, sweeping around, and collecting all the skulking insects that are roused by the trampling of the horses' feet.' Writing at the turn of the nineteenth century, the ornithologist George Montagu in turn reported that 'swallows follow, and repeatedly fly round with great ease, a horse in full trot … in order to pick up the flies roused from the grass.'

It is not always easy to see which insects and other invertebrates swallows are feeding on. Studies have revealed that they take a very wide variety of species, from more than eighty different families, including beetles, wasps, bees, ants, aphids, moths and spiders, as well as flies. What they catch at any particular time depends, of course, on availability. On summer days, swarms of flying ants often emerge en masse, while early in the season flies are the most frequent food, with larger species – such as horseflies and hoverflies – making up a high proportion of the chicks' diet.

As always, the adult swallows must make constant trade-offs between larger prey, which is usually less abundant and harder to catch, and smaller items, which may be easier to find but provide less energy. Given the choice, they tend to favour larger items, which are more nutritious for them and their chicks. Generally, the later in the season, the smaller the individual prey items, though this does vary: for example, moths may be more abundant in summer, when second broods are in the nest.

Early observers debated whether swallows were active or passive hunters of flying insects: did they simply open their beak wide and wait for an insect to fly in, or did they hunt with their beak closed, opening it only to seize their unwary victim? Edward Stanley, the mid-nineteenth-century Bishop of Norwich, was in no doubt how swallows (and other wide-mouthed birds such as the swift and the nightjar) feed:

> [These species] may be said to be almost all composed of mouth, so wide and gaping are their short beaks; consequently when the supply of insects is abundant, they have little more to do than fly with open mouth and close their beaks upon the objects which cross their flight. This the Swallow does with a sharp clicking jerk, which may be heard by an attentive listener on a calm day at a considerable distance.

He was right about the sound of the swallow's bill closing, but wrong about the technique. Writing a few years later, William MacGillivray – a much more thorough and reliable observer – finally cleared up the issue:

> I have carefully watched this species as it flew past ... to discover whether it keeps its mouth open, and I am decidedly of opinion, that in flying, it invariably retains its mandibles in close apposition, until it comes up to an insect. Indeed, the notion of its flying with open mouth is preposterous ... for were swallows to proceed in this manner ... the extreme velocity with which they rush against the air, would necessarily force it into their stomach.

MacGillivray was, of course, quite correct.

William MacGillivray was one of many writers who have noted the swallow's longstanding connection with weather folklore. There are several variations on a rhyme that uses the feeding behaviour of swallows to predict the weather the following day, including:

> Swallows high, staying dry.
> Swallows low, wet will blow.

The link between the height of hunting swallows and the next day's weather goes all the way back to the ancient world, in writings by the third-century BC Greek poet Aratus. In *Diosemeia* ('On Weather Signs', a work largely borrowed from an earlier work by Aristotle) Aratus included more than 400 short verses on weather forecasting. Many were directly related to the behaviour of wild creatures, including the swallow:

> Often when rain is coming ... swallows keep darting over the lake, rippling the surface with their chests.

Another version of the same idea can be found in these lines by the early eighteenth-century poet John Gay:

> When swallows fleet, soar high, and sport in air,
> He told us that the welkin [i.e. the sky] will be clear.

This quotation was used by the lexicographer Dr Johnson in his famous *Dictionary* to support its definition of 'fleet' as 'swift of pace, quick, nimble, active', while in 1828 the scientist Sir Humphrey Davy also connected swallows' feeding patterns with impending changes in the weather: 'Swallows follow the flies and gnats, and flies and gnats usually delight in warm strata of air [i.e. rising thermal air currents].'

But is it (and the many other versions of the notion) true? It is certainly correct to say that swallows, martins and swifts often fly high in the sky on a summer's evening during fine, settled weather, when the warm air has allowed the insects on which these aerial hunters feed to rise higher into the atmosphere. Likewise, when swallows are flying low, it can be an indication that unsettled weather has forced their insect prey to stay closer to the ground. However, they may fly low whatever the weather, as larger flies tend to be found among flowers and around livestock such as cattle.

So, since we can often tell tomorrow's weather simply from whether today's is settled or unsettled, the height of swallows is a long way from being a clear and reliable forecast. MacGillivray unpicked the conundrum with his usual rigour and clarity:

> Many prognostics of the weather are derived from this species; but most of them are erroneous, and some apparently fabricated by persons who have not studied their motions. Swallows fly low or high according to the flight of insects, which is influenced by the state of the atmosphere; but little judgement can be formed of the future by attending to their motions. Thus, they fly low when there is a smart cold breeze, especially if it be accompanied with moisture; but this circumstance is not an indication of the continuance of rain or cold.

One evening last August, as I returned from walking the dog, the silage field behind our home, already cut twice this year, was again showing a healthy crop of lime-green grass. Half-a-dozen swallows were cruising back and forth, a foot or two above the waving sward. On that breezy August evening, the insects were flying low and, even though according to folklore bad weather should have been on its way, the skies appeared settled (and indeed remained so for the next few days).

I always marvel at the way swallows hunt: propelled forward on shallow wingbeats, then twisting momentarily to grab an unseen insect, before moving inexorably on. Some of the birds

above the silage field were young, not long out of the nest: their short, stubby tails and slightly jerkier flight a giveaway. By the time I had walked to the other end of the field they had already gone; off to hunt elsewhere, perhaps, or just rising into the clear blue sky to twitter contentedly to one another on a fine August evening.

Swallows may not necessarily forecast the weather for us, yet the weather does have a significant effect on their own lives. In particular, they have developed strategies to cope with exceptionally hot weather, to which, spending so much time hunting in the open air, they are more vulnerable than most other birds.

Unlike us, birds are unable to sweat, so to lose excess heat they open their mouth as wide as they can and pant, rather like a dog. Swallows will also spread their wings when perched, and extend their legs, in both cases to cool down.

Rain, too, can cause problems, especially when prolonged, or accompanied by strong winds. Swallows are unable to feed during heavy downpours, so seek shelter in barns or other farm buildings. This behaviour may have led to a belief widespread across Europe and Asia that swallows provide protection against lightning strikes. In southern Africa, this superstition is even more explicit: in the Zulu language the swallow is known as *Inkonjane*, the 'lightning bird', although the name also applies to other species, including the hamerkop (a waterbird with a hammer-shaped head).

Swallows are also affected by the weather conditions during the breeding season as a whole. Perhaps not surprisingly, cool, wet summers reduce the availability of flying insects, meaning that fewer chicks successfully fledge.

Combining their two broods, a pair of swallows can raise on average just over five young in a cool and wet year, and just over seven in a warm and sunny one. The young raised in a 'good' year are also likely to be larger, stronger and fitter – more likely to survive the long journey to Africa, and then return north to breed the following spring. But very hot and dry summers are not ideal for insects either, and there is some evidence that these may also reduce breeding success. Typically, swallows thrive in what we used to call the typical British summer: 'two fine days and a thunderstorm', with a mixture of warmth, sunshine and rain. The recent trend towards more unpredictable and extreme weather conditions, a result of the disruption to the world's weather patterns caused by the climate crisis, may not bode well for swallows.

Like other birds, swallows face many other threats. Agricultural pesticides are a double whammy: they kill many of the insects on which the birds depend, and may also accumulate in the birds' bodies, resulting in chronic ill health. Even traffic fumes may be an issue: in the United States swallows nesting close to busy roads have been found to have much higher levels of lead in their system than those breeding in more rural areas.

Swallows also have to cope with parasites – mainly feather lice, mites, flukes and fleas. One of the most common is the tropical fowl mite, which sucks the blood of young swallows, and can be found in huge concentrations: up to 10,000 mites in a single nest. As with any parasite, there is a cost to the birds: chicks with high infestations grow less rapidly, and some may even die. The effects appear to be more serious later in the season, especially in second broods, because parasite numbers tend to rise in summer as the weather warms up.

Mites are also bad news for adult swallows: a serious infestation inhibits the growth of their tail and, as we have seen, males with longer tails tend to be more successful when it comes to mating and raising young. The Danish scientist Anders Møller, who has studied swallows more closely than anyone, discovered that females prefer males that are free from parasites. This is not just because they make more effective parents, but because they are likely to be genetically resistant to infestation, a trait they will then pass down to their young.

But the swallows' habit of returning to the same nest site, and often to the same nest, means that they suffer disproportionately from these freeloading creatures, as the legendary entomologist Miriam Rothschild pointed out in *Fleas, Flukes and Cuckoos*:

> It will have become evident from this brief account, mainly of the fauna of martins' and swallows' nests, that the bird-lovers who carefully preserve their habitations from one year to another also unintentionally preserve the louse-flies, fleas, mites and bugs overwintering as larvae and pupae or hibernating in the nest, which are directly responsible for bringing hours of pain, misery, disease or even death to the nestlings in the following spring.

In a letter to the journal *Nature* published in July 1885, William Watts wrote that if a swallow's nest is taken down and examined as soon as the young have fledged, the lining will be 'swarming with two species of active insects altogether out of proportion as to size of the swallow on which they are parasitic'. He also delivered a withering response to the bizarre suggestion by a previous correspondent that these creatures were deliberately encouraged by the nesting swallows to provide food during their long journey south – a kind of portable picnic: 'In my opinion there is no design or intention on the part of the swallow to breed or cultivate parasites for consumption during migration. The life of the parasite depends on the existence of the swallow, and not the swallow upon the parasite.'

Not all unwelcome guests are found in nests, however. Swallows are also one of the main hosts of the bird malaria parasite, which Rothschild suggested was acquired in autumn, when the birds roost together in reedbeds before heading off on their journey south.

But at least they are spared the fate of another migrant, the swift, which carries a cohort of feather lice all the way to Africa and back. This is the equivalent, as Miriam Rothschild memorably noted, of a human traveller living with 'a couple of large shore crabs scuttling about in his underclothes'. Not a pleasant thought.

The summer months bring another, more immediate danger for young swallows. Less experienced and far less manoeuvrable than their parents, they can be easy targets for any aerial predators, as Gilbert White observed:

> About the tenth of July in the same summer a pair of sparrowhawks bred in an old crow's nest on a low beech in the same hanger ... the old birds had been observed to make sad havoc for some days among the new-flown swallows and martins, which, being but lately out of their nests, had not acquired those powers and command of wing that enable them, when more mature, to set such enemies at defiance.

W. H. Hudson pointed out that the swallow does have a defence mechanism, acting as a sentinel for all birds in danger of being seized by the sparrowhawk:

Spar-row-hawk.

> The swallow has ... a loud, startled, double alarm-note, uttered at the appearance of a hawk speeding through the air ... The appearance of a hawk excites as much anger as fear, and he generally goes in pursuit of it; but the note is understood by other small birds, and has the effect of sending them quickly into hiding.

John Clare also noted how swallows take a lead role in drawing attention to danger: 'Small birds have as strong an antipathy against hawks particularly the swallows & martins who will flock round one as he passes from wood to wood ... uttering a twittering note.'

Once I had been living in Somerset for a year or two, I began to get a sixth sense for when the local sparrowhawk was hunting newly fledged swallows. In those days, the barns next door had not yet been converted into homes, and several pairs of swallows still nested there. Mowing the lawn or simply relaxing in the sunshine, I became so accustomed to what William Yarrell called the 'soft and sweet warble' of the swallows that I hardly noticed it.

Until, that is, the tone and texture of the notes changed. The usual gentle twittering became a loud, urgent warning. If I looked up, I soon came to realise, in a second or two a purposeful shape with rounded wings and a long tail would appear low over the barns, heading directly for the family parties of swallows. More often than not, the warning calls were enough to alert the young swallows and keep them safe from the sparrowhawk; but no doubt some did fall victim to this low-flying predator.

One hot afternoon in July or August, I became aware of a series of more urgent calls uttered by the adult swallows. Sure enough, the threat this time came from an even more lethal predator: a hobby. This slender falcon specialises in hunting dragonflies in spring, and young swallows and martins during the summer and early autumn, before it follows them south, all the way to Africa.

I was surprised to see the adult swallows not fleeing away from danger, but actively courting it: chasing the hunting hobby from behind and mobbing it angrily to drive it away. This is a risky business: the only time I have seen a hobby seize a small bird – an unwary house martin, feeding over Staines Reservoirs in West London – the impact was as loud and forceful as the crack of a whip.

A few miles east of where I live, the road that winds from Wedmore towards the cathedral city of Wells passes through the village of Latcham. On the left-hand side, a sea of cars await repair at the local firm of Ratcliffe Brothers. With mild nostalgia I parked my car next to a Triumph Stag, its registration plate showing it dated back to the late 1970s. But I was not here to look at cars, but at birds.

Above the clunks and banging of cars being repaired, I could hear the familiar twittering of swallows. At the doors of the workshop, one swooped low over my head, so close I felt a light gust of air from its feathers. Inside, I found the familiar semi-organised chaos of any such establishment: tyres piled up, tools dangling from hooks, Haynes manuals, cardboard boxes, girlie calendars and, of course, cars – each being carefully attended to by expert mechanics, their hammering accompanied by similar rhythms on Radio 2's *Steve Wright Show*.

Over my head, though, was a more unusual sight: a score of large, colourful umbrellas, hanging upside-down and open from the corrugated iron roof. They were marked with various company logos, and also spattered, on the inside, with large amounts of swallow poo. That's because each umbrella had been carefully placed underneath a swallow's nest, to protect the cars – and mechanics – beneath.

Then I noticed them: four baby swallows, perched along two metal struts a foot or two below the roof. Young swallows have little of their parents' sleek elegance. They share the same basic shape and pattern – a slender, elongated body, dark wings and back, pale underparts and a brick-red throat – but they are much duller than the adults, while their shorter tail gives them a rather stunted, helpless appearance. I could see that one of the four was more active and restless than its siblings, continually shuffling up and down the metal strut, turning and fidgeting, and occasionally flapping its wings as if testing that they were working, while the others sat patiently alongside.

These young swallows would have hatched out several weeks earlier, around the start of July, and only left the safety of their nest a few days ago. They were still very youthful, but each morning and evening they headed outside the workshop to exercise their wings and practise flying and feeding, though they still mostly relied on their parents for food. Every few minutes, one of the adults returned with a tightly packed bolus of insects, provoking the youngsters to open their bills wide and reveal the bright yellow gape. The whole process took barely a few seconds, the adult in and out in the blink of an eye.

This is a family business, the garage's co-owner, Andrew, told me as I watched: established by his father, who had moved from Manchester to Somerset back in the mid-1960s. Raised locally, Andrew and his younger brother Duncan took it over in the 1970s. For the first few years there had been no swallows breeding here. But some time in the mid-1980s, one pair entered the workshop, liked what they saw, and built a nest. Year by year, numbers crept up, until a few years ago they had reached a record total of twenty-one nests. This year numbers were well down, said Andrew: just five first broods and seven second ones, compared with eighteen the previous breeding season. He suggested that this might be because last year's 'drought summer' had been just too dry – fewer insects for the swallows to feed their young.

This summer, his hope is that some of the birds whose second broods are already out of the nest and beginning to gain their independence will have a third before they finally headed south. But this is a real gamble: on several occasions Andrew has seen baby swallows left behind at the end of September, still not ready to migrate, and most likely doomed to die here or en route.

The key to the birds' success at the garage is that they are safe: protected by the presence of people. Swallows' nests are very vulnerable to predators: cats and sparrowhawks by day, owls by night. In the early years, after the doors had been closed for the night, some birds of prey did find their way inside, and wreaked havoc. To stop the carnage, the brothers replaced the window through which the predators were getting in with a tiny, letter-box-like slit – about a foot wide and barely two or three inches deep – through which the swallows could still fly. Even so, the birds are not always safe: babies often fall out of the nest and have to be put back.

The brothers recall their late father's delight when the birds would return each April – 'They're here, they're here!' – while their mother, who still lives just down the road in Wedmore, eagerly awaits the news of their arrival. Andrew shares her joy: 'Every April when the first one appears, I get so excited, and have to ring my wife to let her know they're back!'

The customers, many of whom have been coming here for decades, love it too; regulars often ask about how the birds are faring; some even bring in more umbrellas to hang up. Andrew admits that the mechanics are not always happy, 'especially when the swallows crap in their toolboxes', but they have learned to tolerate these charismatic birds.

For Andrew and Duncan, the presence of these birds is something very special. As Duncan says, 'They're a portent of something better to come: the swallows are here, the weather's getting better, and we've got the whole summer to look forward to.'

In this quiet Somerset village, humans and swallows have entered a very special co-existence: a kind of symbiosis that must go back in some form at least to medieval times, when farmers built the first barns to store their grain, and inadvertently provided a new home for nesting swallows.

In the skies above the car repair workshop – indeed, the skies all over Somerset – fine afternoons see the swallows out in force grabbing as many insects as they can to feed their hungry broods. They look as if they are flying at great speed – especially when they skim low over the ground – but this is something of an illusion. Typically, when feeding, swallows fly at between 8 and 11 metres per second – 18–24 miles an hour, or roughly the pace of a racing cyclist. This does of course vary: when pursuing a fast-flying item of prey they may reach speeds of almost 19 metres per second, or over 42 miles per hour – far quicker than even the fastest human sprinter. William Shakespeare was much taken with the swallow's rapid flight: in a speech by the Earl of Richmond in *Richard III*, he conjured up this delightful image to foretell the humble Richmond's elevation to the throne as Henry VII:

> True hope is swift, and flies with swallow's wings.
> Kings it makes gods, and meaner creatures kings.

One thing that separates swallows and martins from most other songbirds is their flight efficiency. Typically, they use only 60 per cent as much energy as other species when airborne: a huge advantage both when feeding and migrating. It is also a necessity, given that they take virtually all their prey on the wing.

The distinctive shape of all hirundines is a result of this need to be energy-efficient: their bodies are long, narrow and streamlined, while their wings are triangular – rather like the delta-wing design used on supersonic aircraft. Compared with other birds, they have a very low 'wing loading' (the bird's body weight divided by its wing area). The swallow's tail is also crucial: it serves as a rudder, allowing the bird to apply drag or lift, and so change direction rapidly in mid-air.

The swallow's unusual wing shape enables it to fly not just fast but also very slowly, and to twist and turn to chase and catch its target. Even compared with swifts and house martins, swallows are the most manoeuvrable of all the aerial feeders, able to turn through a right-angle in less than their own body length. No wonder that distinctive method of try-scoring in rugby union, in which the player leaps over the line to evade his opponent's tackle, is widely known as the 'swallow dive'.

This also explains why swallows are such a joy to watch, as Thomas Bewick so beautifully observed:

> The Swallow lives almost constantly in the air, and performs many of its functions in that element; and whether it pursues the devious windings of the insects on which it feeds, or endeavours to escape the birds of prey by the quickness of its motion, it describes lines so mutable, so interwoven, and so confused, that they can hardly be pictured by words.

After a couple of rather wet and windy weeks, the end of August produced a spell of sunny, warm and settled weather, just in time for the bank holiday. An ideal opportunity to mow the lawn. Being rather lazy, however, I had embarked on a new regime. Instead of collecting the grass cuttings in a wheelbarrow and pushing them all the way down to the bottom of the garden to our compost heap, I removed the grass box and simply allowed the grass to remain on the lawn – my own small-scale version of haymaking.

This clearly appealed to the swallows. Shortly after I had finished, and was sitting on our veranda enjoying a cold beer, they came down: a dozen or more, swooping over the newly mown lawn to grab a fast-food bonanza of low-flying insects. As usual when feeding, they twittered companionably to one another; now and then, a small group flew up into the sky and called more insistently, an amenable, cheery sound I always miss when they leave.

Writing a book about the swallow has, inevitably, led me to notice them wherever I go. The same was true when I was working on the first book in this series, *The Robin*: that summer my garden seemed full of baby robins. Writing the second book, *The Wren*, during the early months of the year, seemed to encourage wrens to pop up in the most unlikely circumstances – on one memorable occasion, a male singing on the balcony outside our bedroom window as my wife and I lay in bed.

This summer, almost as if they know I am writing about them, swallows have been everywhere: filling the Somerset skies with their sounds and movements. Skimming and swooping, skidding and flouncing, plunging and cruising – I'm running out of words to describe them.

I marvel at their ability to keep moving: turning the insects they catch into this constant expenditure of energy like some avian perpetual-motion machine. *Nyankalema* – the name given to the barn swallow in Zambia – means 'the bird that never gets tired', for these birds are incessantly active during the hours of daylight.

At times – especially when they join forces with their fellows and launch themselves into the stratosphere – they really do seem to be enjoying themselves. Nature writers often frown on anthropomorphism (though few of us can avoid indulging in it from time to time), but it really is very hard to write about the swallow (and indeed about the robin and the wren) without occasionally comparing their lives to our own.

In 1996, towards the end of a distinguished career as a field ornithologist spanning more than seven decades, Alexander Skutch wrote a fascinating book, *The Minds of Birds*. Birds, he postulated, do not always follow pure biological instinct in their day-to-day behaviour, but sometimes indulge in what can only be described as 'play'. Skutch told the story of the mid-twentieth-century, Devon-based naturalist and musician Gwendolen Howard, who wrote books and articles under the masculine pseudonym 'Len'. She described watching swallows as they flew low over a grassy Devonshire hillside where the local ducks and geese had been shedding their feathers, which then floated in the summer's breeze:

> A swallow would drop down, seize a feather in its bill, then swoop upward to circle above the other swallows and drop the plume.

As it floated down, the other caught it. So the game continued while with marvellous grace the birds traced wide arcs through the air.

When I watch swallows riding the air currents over my garden on a warm, late-August evening, I find it hard to disagree with Skutch and Howard's belief that these birds are simply doing it for fun (though of course it may also help the youngsters hone their flying and hunting skills). Certainly, I feel joy watching them; a joy tinged with a sweet sadness that, within a month or so, they will have flown away on their perilous journey halfway across the world.

Down on the Somerset coast, by early August there were signs that summer was reaching its zenith. In the sheltered corner of grass by Huntspill Sluice, I came across a host of late-season butterflies: bright common blues, a piebald marbled white, the delicate brown argus, showing off the tiny orange spots around the edges of its wings, and the first small heaths I had ever seen there. All were enjoying the best butterfly summer for a decade, the settled, warm weather providing plenty of nectar for them to feed.

But nearby, down by the River Parrett, the season seemed to have already shifted from summer to autumn. The first sound I heard was the distinctive '*sweeeep*' of a yellow wagtail, one of several feeding alongside their pied cousins on damp grass recently revealed by the falling tide. Some swallows and a lone sand martin passed overhead on their way south while, further along the sea wall, two wheatears – my first of the season – flashed their white rumps as they flitted along the path a short distance ahead.

In the natural world, the shift from summer to autumn is not a linear process. Rather, it is a Venn diagram in which the two seasons overlap – a phenomenon sometimes demonstrated by one and the same species. So the swallows I had been seeing along the coast may already have been migrating, while those a mile or so inland, by the bridge across the Huntspill River, were still very much in breeding mode. Three of them were perched on a telegraph wire: two adults and their single offspring, begging its parents for food. Close up, I could see the youngster's dirty-brown underparts and yellow line along the edge of its bill – clear signs of its youth. Then all three birds dropped off the wire and, twittering as they went, swooped down over the river to feed, their deep-blue plumage glittering in the midday sun. These swallows wouldn't even think about migrating for a month or more; whereas the yellow wagtails and wheatears – neither of which breed nearby – were clearly passage migrants: signs that autumn was already underway.

On the penultimate day of August, my family and I went on holiday with my in-laws in the heart of West Sussex. The moment we arrived at our rented home, I could hear the twittering of swallows, almost as if they had followed us here from Somerset. Of course, they didn't need to: wherever I travel in rural lowland Britain, from Somerset to Sussex and Scilly to Shetland, I see and hear these captivating birds – at least during the half of the year when they are visiting the UK.

I had noticed a swallow swooping low outside the garage, but

it wasn't until a day or two later that I realised they were still nesting there, when my young nieces came running in, full of excitement, to tell me the news. At the back of the garage, to my great delight, I found no fewer than four newly fledged swallows, all lined up along a shelving unit next to a poo-spattered ladder, each looking straight back at me. They had the classic appearance of all baby birds: smaller and fluffier than their pristine, elegant parents, and with the slightly baffled gaze of fledglings confronted with their new and presumably rather frightening world.

The nest itself was attached precariously, as they so often are, to the side of one of the wooden beams beneath the low ceiling. Swallows usually build on top of a beam, but this one was constructed more like that of a house martin. As always with swallows' nests, though, the bits of straw sticking out of the structure made it look far more slipshod than the neat nest of their smaller cousin. The parents came back, twittering loudly as if to encourage their recalcitrant brood to venture outdoors; which two of them did, leaving their less-daring siblings behind. Then, as I went to leave, the others finally took to the wing, and shot outside.

That nest had done well to survive hosting this large and very active quartet. Not all breeding swallows are so fortunate: in the village next to ours, at the home of some friends, the nest in their front porch fell down while they were on holiday. When they returned, the baby birds were lying dead on the floor.

This is a perennial problem for nesting swallows. More than two centuries ago, the writer Dorothy Wordsworth noted in her diary for 25 June, 1802,

> When I rose I went just before tea into the Garden. I looked up at my Swallow's nest and it was gone. It had fallen down. Poor little creatures they could not themselves be more distressed than I was ... They lay in a large heap upon the window ledge ...

Four days later, however, Wordsworth had better news: the parents had rebuilt the nest, though whether their fallen brood ultimately managed to fledge, she does not say.

If they do manage to avoid such calamities, swallows usually raise two broods, especially in fine summers. Spring weather is also important: the earlier a pair of swallows gets down to laying their first clutch of eggs in April, the more likely they'll be able to raise another brood later in the season. It takes roughly seven weeks to rear each brood (from the first egg being laid to the last chick no longer needing to be fed), and then the female needs time to recover her energy levels, so squeezing in a third brood is touch-and-go.

Birds that return in late March and get straight down to breeding might be able to have fledged chicks by early May, and a second brood of fledged chicks, say, ten weeks later, in late July. If the sun is still shining, and there is still plenty of food, they sometimes have another go: if they lay a third clutch of eggs in early-to-mid-August, and these hatch by early September, they might just pull it off. But it only takes a turn in the weather, with September gales sweeping in from the Atlantic, to drastically reduce the amount of food available for those late youngsters.

Like many insectivorous birds, including most warblers, flycatchers and chats, swallows in Britain and Europe migrate south because they have to. Although there will still be an abundance of insects well into September, the shift in weather patterns coinciding with the autumn equinox soon leads to a rapid reduction in the numbers and density of prey. That's particularly true for swallows, martins and swifts, because they feed almost solely on prey caught in flight. While small insects remain on foliage (some birds that feed on these, including the goldcrest, stay put for the whole winter), from October all the way through to March flying insects are few and far between. Add in the shortening hours of daylight, and it's easy to see why swallows head away.

Some go much earlier. The fledged juveniles from the first broods – which I have seen as early as the second week of May – may leave in July, sometimes followed by adult birds that, for whatever reason, have not begun a second family. They arrive in South Africa in mid-September – the equivalent of early spring – so spend well over half their lives away from Britain. The main autumn migration takes place from late August through to early October, but may extend into November, with a few stragglers leaving as late as December.

So, unlike swifts and cuckoos, which seem to be here one minute and gone the next, the swallows' departure is often long and drawn-out. They seem to revel in the planning stage, spending weeks in the avian equivalent of the departure lounge, before finally heading off.

But how do birds know, asks the late Sandy Denny in that tender and plangent song 'Who Knows Where the Time Goes?', that it's time to go? The reason is that, as the Earth slowly shifts on its axis from north to south, marked by the autumn equinox, subtle changes in their brain trigger the impulse to leave us for warmer climes.

Almost a quarter of a century ago, I experienced what remains my most memorable encounter with departing swallows. It was during the annual British Birdwatching Fair – 'Birdfair', as it is usually known – held on the third weekend of August at Rutland Water, in England's smallest county, where I was filming a TV programme for the very first series of *Birding with Bill Oddie*.

One morning, as we drove through the picturesque little

village of Egleton, we noticed a number of slender, long-tailed birds gathering on the telegraph wires just a few feet above our heads. Looking more closely, we realised we had hit the jackpot: not just swallows, but house martins and sand martins, too; all lined up alongside one another in a rare display of mutual harmony.

Later in the season, they might have been more agitated, as the migration instinct began to kick in. But with several weeks to go before they would even think of setting off south, they had stopped for a few minutes to preen, running their beaks through their wing feathers to be ready to fly and feed again. When they fluffed up their plumage, observed Bill, they managed to look both cuddly and sleek at the same time.

The four parallel wires reminded me of a musical stave and, when we returned to base with our footage, Ed Bazalgette, our creative and talented editor, thought so too, and added a soundtrack of a pizzicato guitar and violins.

Bill and I were usually very reticent about using music to illustrate our programmes, because to us it broke the spell we were trying to create – the sense of the viewer actually being there, with Bill, simply looking at the birds.

But in this case, it really did work, with each plucked note and the fast-paced rhythm perfectly capturing the restless spirit of these delightful little birds as they fluttered to and fro. Bill's commentary ended with a suitably valedictory farewell: 'Three different species on the wires: all very endearing, all very attractive, all migrants, and in a few weeks from now – all gone!'

4

AUTUMN

Everything about swallows says 'South'.

Roger Deakin, 'Follow the Swallows'
(posthumously published, 2015)

During the first week of September, perfectly timed for the start of the new school year, the weather began to turn. For the first time since April, swallows were few and far between. Towering, anvil-shaped clouds and a chilly north-westerly breeze told me what I already knew: autumn had arrived.

Out with Rosie on an early-evening walk, I finally connected with a handful of swallows, powering low over the hedgerows. Heading south? Maybe not quite yet – there would still be sunny, warm, insect-filled days between now and the end of September – but their flight certainly had a purposeful feel.

This summer had been warmer and drier than most. Yet the older I get, the more quickly the once slow and languid days of summer seem to pass. The swallows' twittering now also felt as if it had an urgency, as they sought out the last few insects. Some would still be feeding not just themselves, but a third and final brood of young, in a race for them to fully fledge and be ready to fly south to Africa before the weather turned too cold.

Three weeks later, Radio 4's *Today* programme reminded me that it was the first day of autumn. From now on throughout these northerly latitudes, the days would be shorter than the nights. Autumn arrived in Somerset with a vengeance: after a warm and settled few weeks, a front swept in from the Atlantic, bringing cloud, murk and a constant drizzle. Bad news for my county cricket team, set to lose their outside chance of winning the Championship because rain had stopped play, and not great news for the single swallow that passed over my head, doggedly heading into the stiff south-westerly breeze like a scrap of paper, on the start of its journey south.

Later that day, once the weather improved, I watched a swallow perched on a telegraph wire by a farm at the edge of the village. She could only have been out of the nest for a few weeks, for her tell-tale short tail and duller colours still marked her out as a young bird. She would have fledged at the height of the summer heatwave, but would now be sensing an unfamiliar chill in the air. I imagined the taut wire buzzing beneath her tiny feet, a human voice being converted into an electrical signal and sent elsewhere: just one of hundreds of millions of signals criss-crossing the nation on this and every other day. In this swallow's tiny brain, there would be other signals, as the daily diminution of daylight caused the release of chemicals encouraging an unstoppable wanderlust. It was almost time to head off.

She was not alone. Her surviving siblings, her parents, and the other swallows that nested in the barns and outbuildings of this Somerset farm, were looking restless too. From time to time, they responded to some unseen signal by releasing their grip on the wire and swooping down, twittering to one another in what the poet Rosamund Marriott Watson called their 'sweet-voiced travel talk'. Each time, I wondered if they would return to their wire, or if this was it – for I knew that very soon they would disappear from the skies above altogether, and not return for six months or more.

Momentarily I was filled with wonder and awe. Once they did finally leave, she and millions of others would head across the English Channel to France, cross the Mediterranean at its narrowest point from Gibraltar to North Africa, and then face the unimaginable vastness of the Sahara, where many of her companions would meet their death.

If she survived, she would cross the Equator, continue south for several thousand miles, until she finally reached the Limpopo River where, unbeknownst to her, at the start of the year her father had set off on his journey north. She would face the same challenges he did: predators, unexpected storms or heavy rainfall, or simply a lack of insect food. If she overcame them all, she would spend the winter with millions of other barn swallows from all over Britain, Europe and western Asia, in southern Africa, until she flew back home again the following spring.

'They are a shiny, metallic, gregarious, nomadic tribe,' wrote Roger Deakin in his wonderful essay, 'Follow the Swallows',

> decked in magenta and ravishing deep blues like the Tuareg and Bedouin, whose deserts they must cross as they set out in September … As they gather talkatively on the telegraph wires, they seem no more afraid of the great distances they must travel and the hardships they will encounter on the way, than you or I making a long-distance phone call.

How that little bird on the wire, along with the hundreds of millions of other migratory birds, finds her way there has puzzled lay people and poets, curious observers and professional ornithologists, for millennia. It must work, otherwise these long-distance migrants would have died out long ago. But look at it from their point of view and migration – despite its many risks – gives the best of both worlds. Indeed, the Swedish migration scientist Thomas Alerstam turned the whole conundrum on its head:

> The question is often asked, 'Why do birds migrate?' ... [Yet] birds and migration belong inseparably together. Rather, we should be surprised that there are some bird species that are very sedentary: why do not all birds migrate?

That may be the prevailing wisdom today, but for centuries the notion that birds like the swallow left our shores each autumn, and travelled thousands of miles to spend the winter in the Southern Hemisphere, before returning north the following spring, was regarded as utterly incredible. This led to some of the most bizarre pseudo-scientific theories ever propounded.

The sixteenth-century Swedish scholar Olaus Magnus was, by all accounts, a very learned man. The occasional depiction of a sea monster aside, he thrilled the educated world with vivid accounts of Scandinavian myths, folklore and customs, which were first published in Latin, and a century later translated into English.

On the title page of his 1555 work, *A Description of the Northern Peoples*, he set out the wide range of subjects it covered, including 'the wonderful differences in customs, holy practices, superstitions, bodily exercises, government and food keeping ... war, buildings and wonderful aids ... metals and different kinds of animals, that live in these neighbourhoods'. And there was his theory on the migration of swallows:

> Although the writers of many natural things have recorded that swallows change their stations, going, when winter cometh, into hotter countries; yet, in the northern waters, fishermen oftentimes by chance draw up in their nets an abundance of swallows, hanging together like a conglomerated mass.

Olaus Magnus got many things right, despite the limited knowledge of natural history back in the sixteenth century, but on bird migration he was spectacularly wide of the mark. He did grudgingly accept that swallows might head a little way south from northern Europe to find warmer climes but, like so many, simply could not credit these tiny creatures travelling such vast distances across the globe. He even described exactly how swallows achieved this submersive hibernation: 'In the beginning of autumn, they assemble together among the reeds; where, allowing themselves to sink into the water, they join bill to bill, wing to wing, and foot to foot.' A striking image, if a rather Pythonesque one.

Olaus Magnus was not alone. Gilbert White, often hailed as the first modern nature writer, returned to the subject of swallow hibernation no fewer than eight times in *The Natural History of Selborne*. The first entry, for 4 August 1767, does show a degree of scepticism: 'As to swallows being found in a torpid state during the winter in the Isle of Wight, or any part of this country, I never heard any such account worth attending to.' Yet he then goes on to relate two tantalisingly convincing – though second-hand – accounts of just such behaviour:

> But a clergyman, of an inquisitive turn, assures me that, when he was a great boy, some workmen, in pulling down the battlements of a church tower early in the spring, found two or three swifts ... which were, at first appearance, dead, but, on being carried toward the fire, revived ... Another intelligent person has informed me that, while he was a schoolboy at Brighthelmstone, in Sussex, a great fragment of the chalk-cliff fell down one stormy winter on the beach; and that many people found swallows among the rubbish; but, on my questioning him whether he saw any of those birds himself, to my no small disappointment, he answered me in the negative; but that others assured him they did.

In February 1769, White described an incident that had taken place the previous Michaelmas Day, 29 September 1768, when he had left Selborne early in the morning, amid 'a vast fog'. By the time he was a few miles from home, the sun finally came out, and he and his companions noticed 'great numbers of swallows' gathered on low shrubs and bushes. White correctly presumed that they had roosted there overnight, before flying off the following morning. Yet, even having witnessed the birds heading south for himself, he went on to repeat the same error, asserting that some swallows 'never leave this island, [but instead] lay themselves up, and come forth in a warm day, as bats do'. He supported this view with yet another unverifiable observation:

> For a very respectable gentleman assured me that, as he was walking with some friends under Merton [Oxford] wall on a remarkably hot noon, either in the last week of December or the first week in January, he espied three or four swallows huddled together on the moulding of one of the windows of that college.

What is odd about White's refusal to accept the notion of bird migration is that his younger brother John was stationed in Gibraltar, where he served as chaplain to the British forces. Also a keen observer of nature, John regularly sent his elder sibling reports of his observations at this migration hotspot, writing that 'myriads of the swallow kind traverse the Straits from north to south.'

Situated where the distance between the continents of Europe and Africa is at its shortest, Gibraltar sees a twice-annual concentration of many thousands of migrants, as even Gilbert White himself noted:

> My brother has always found that some of his birds, *and particularly the swallow kind* [my italics], are very sparing of their pains in crossing the Mediterranean; for when arrived at Gibraltar, they scout and hurry along in little detached parties of six or seven in a company; and sweeping low, just over the surface of the water, direct their course to the opposite continent at the narrowest passage they can find.

A few years later, though, in a letter to the eminent scholar Daines Barrington, White had retreated to the default position of his day:

> Dear Sir – You are, I know, no great friend to migration; and the well-attested accounts from various parts of the kingdom seem to justify you in your suspicions, that at least many of the swallow kind do not leave us in the winter, but lay themselves up like insects and bats, in a torpid state, to slumber away the more uncomfortable months till the return of the sun and fine weather awakens them.

The same year, he used a first-hand sighting of tardy autumn swallows to confirm his mistaken belief:

> As a gentleman and myself were walking on the fourth of last November round the sea-banks at Newhaven … we were surprised to

> see three house-swallows gliding very swiftly by us. That morning was rather chilly, with the wind at north-west; but the tenor of the weather for some time had been delicate, and the noons remarkably warm. From this incident, and from repeated accounts which I meet with, I am more and more induced to believe that many of the swallow kind do not depart from this island; but lay themselves up in holes and caverns; and so, insect-like and bat-like, come forth at mild times.

In his final entry on the subject, he reasoned that, during late cold spells in spring, when newly arrived swallows disappear, they have retreated to their 'hybernaculum' rather than heading back south again.

It is tempting to give White the benefit of the doubt, yet the phenomenon of bird migration is mentioned by writers in

Ancient Greece and Rome, and even appears in the Old Testament Book of Jeremiah – 'Yea, the stork in the heaven knoweth her appointed times; and the turtle [dove] and the crane and the swallow observe the time of their coming' – so it does seem odd that he clung onto these mistaken theories for so long. To be fair, though, it must have been difficult for a parish priest, confined to a quiet country village and its immediate environs, to even begin to imagine the reality of a swallow's life when it heads away from our shores.

Hibernation may seem the ultimate in weird and wacky theories, but another explanation of the annual disappearance of swallows manages to make it look almost sensible. Towards the end of the seventeenth century – just before what would come to be called the Age of Enlightenment – an Oxford scholar named Charles Morton came out with an even more outlandish theory: that swallows did migrate – not to Africa, but to the Moon.

His reasoning went as follows: swallows could not possibly hibernate at the bottom of ponds or lakes, as the water would be too cold, and they would die from a lack of oxygen. The Moon, on the other hand, would provide the ideal home for these birds, albeit a rather distant one. Morton's estimate of the mileage involved – a round trip of just under 360,000 miles (roughly 575,000 kilometres) – was not far off the actual distance of between 450,000 and 500,000 miles (720,000–800,000 kilometres). Having done the maths, he justified his hypothesis with a wonderfully circular piece of reasoning: 'Now, whither should these creatures go, unless it were to the Moon?'

Not all his contemporaries were quite so credulous. The

pioneering naturalist John Ray, writing in the 1670s, asserted that 'it seems probable that they fly away into hot countries', while, more than a century later, Thomas Bewick wrote of experiments carried out by a Mr James Pearson of London. At the end of August 1784, Pearson had taken five or six swallows into captivity (captured using a net), placing them together in a cage. The first winter, they mostly died, but the following year he repeated the experiment with greater success: all the birds survived, singing throughout the winter, and then moulting into their full breeding plumage – proving that they did not go into a state of torpor or hibernation in the winter months. Instead, as Bewick correctly concluded: 'They leave us when this country can no longer furnish them with a supply of their proper and natural food; but more especially, when the great object of their coming, that of propagating their kind, has been fulfilled.'

Bewick also reported sightings in spring, from around the Balearic Islands in the western Mediterranean, of large numbers of swallows flying northwards. And in his *Ornithological Dictionary*, published in 1802, Bewick's contemporary George Montagu gave short shrift to the hibernation theory:

> Why it should be necessary to account for the loss of this tribe of birds in the winter, by making them to immerse during that season, is extraordinary, when at the same time no doubts have been entertained of the migration of other birds, whose powers of wing are far inferior. And yet there have not been wanting persons who have declared they have seen them drawn up in nets, and restored from their benumbed state.

Despite all the evidence to the contrary, however, White's error was repeated for almost a century after his death. As late as 1865, the woefully inaccurate popular nature writer Bishop Edward Stanley continued to spread the myth of hibernation, though he did hedge his bets:

> We will not positively assert that Swallows can … continue through the winter in a dormant state, and still less, that they can exist at the bottom of water; but as instances well attested, without discernible reasons for deceiving, are abundant … they at all events merit some notice.

Soon afterwards the hibernation myth was finally debunked, and in 1866, in *Birds of Middlesex*, J. E. Harting sensibly hypothesised that 'It is not unlikely that this propensity to roost among willows and near water has given rise to the popular delusion that Swallows retire under water in winter.'

The close observation John Clare brought to nature taught him exactly what swallows did in autumn, and he too was aware of them gathering in large numbers over water before departing south:

> They generally haunt rivers & brooks before they start [migration] & may be seen settling 4 or 5 together on twigs of Osiers beside the stream that bend with them till they nearly touch the water – they make westward when they start & often return agen resting by flocks on churches & trees in the village as if they were making attempts before they started for good.

One of his finest short poems, 'On Seeing Two Swallows Late in October', opens with a bittersweet evocation of the final swallows of the autumn:

> Lone occupiers of a naked sky
> When desolate November hovers nigh
> And all your fellow tribes in many crowds
> Have left the village with the autumn clouds …

Then, in sentiments we can all recognise stirred up by this transitioning time of year, he anticipates their long journey ahead 'to the farthest lands / oer untraced oceans and untrodden sands', and yearns for these 'little lingerers' to stay put for the winter:

> I wish ye well to find a dwelling here
> For in the unsocial weather ye would fling
> Gleanings of comfort through the winter wide
> Twittering as wont above the old fireside
> And cheat the surly winter into spring.

So prolonged and contingent is the annual departure of the swallows – and sand and house martins – that I am never quite sure when I have seen my last for the year. The latest I have ever seen a swallow in my home village was 12 October, which, with appropriate symmetry, was the same day the first redwing arrived from Iceland to spend the winter here. Such are the occasional unlikely meetings of migrants from different parts of the globe.

Last year, I was determined to get one more 'swallow fix' to sustain me for the winter, and where better than the south coast? So it was that in early October, on a blustery but sunny day, I snaked my way through the traffic jams, quarry lorries and a few late holidaymakers' cars onto the Isle of Portland in Dorset, and finally emerged on the open road towards 'the Bill'.

Portland is, it must be said, a rather strange place. It is what geographers call a 'tied island': not quite separated from the rest of Dorset, but linked only by a narrow isthmus – more properly, a tombolo – of land. Like many true islanders, those born and bred here have a slang word for incomers: 'kimberlins'. The main industry is the quarrying of the famous Portland stone, the durable limestone long used in the capital for building, for example, the Tower of London, the British Museum, St Paul's Cathedral and Broadcasting House, home of the BBC. Thomas Hardy, the best-known of all Dorset's authors, described Portland as 'the peninsula carved by Time out of a single stone', which sums the place up pretty well. Portland locals are also obscurely superstitious about rabbits – which they steadfastly refuse to name, referring to them as 'underground mutton', 'long-eared furry things' or simply 'bunnies'. One theory is that quarry workers loathed the animals because their burrowing caused rock falls.

I was here to see not bunnies, but birds: specifically, swallows. The last had flown through my village a week or so ago, but I suspected there would still be a few here. My destination was the Lower Lighthouse: a Grade II-listed building originally opened in the nineteenth century, but which became derelict during the Second World War. In 1961, thanks to a generous donation by a local conservationist, Helen Brotherton, it was reopened as a bird observatory. Today, Portland Bird Observatory and Field Centre – known simply as the 'Obs' – is the most southerly of nineteen such places in Britain and Ireland, found mostly on coastal headlands and remote offshore islands.

It may have been unseasonably warm, but it was also very windy – and getting windier, according to the forecast. Out at sea white horses were forming on the crest of each incoming wave. Gulls and gannets streamed steadily westwards, the latter undeterred by the strength of the wind, which they ploughed through on their stiff, white, black-tipped wings. I sought shelter in the Obs, where two visiting birders kindly offered me a cup of tea and a slice of home-made coffee-and-walnut cake, while I waited for the man I had come to see: the warden, Martin Cade.

Tanned and weather-beaten, Martin is almost as permanent

a fixture as the lighthouse himself, having worked here for the past thirty years. When I asked him if he ever went to Birdfair, he looked at me with a wry smile: 'Why would I want to go and stand in a tent with a bunch of birders when I can be here, enjoying the birds!' His passion for Portland has certainly borne results. The official list of bird species recorded here now stands at just over 300, of which Martin has seen almost all, probably giving him the highest 'patch list' of anyone in the country. As we drank our tea on the observatory terrace, he pointed out the area in front of us, a well-vegetated garden running down to the sea, as probably the best single place in Britain to watch birds. Since its rivals – Fair Isle, North Norfolk and Scilly – cover much larger areas of land, he may well be right.

Not that there was an awful lot to see today. The regular kestrel was hovering in the stiff breeze; a few flocks of starlings, linnets and goldfinches were passing overhead. Then, what I was looking for – a couple of swallows flying into the westerly wind.

Martin swiftly disabused me of the notion that these birds were about to launch themselves out to sea. From here to the French coast at Cherbourg, he pointed out, was 70 miles, so most swallows will head eastwards along the coast to places like Dungeness, from where they can almost see the other side of the Channel. 'It's like they get to the end of the Bill and think, "Bugger that, we're not going any further!" and then turn around.' It was more likely they were coming here, he explained, 'to get the lie of the land', before their final embarkation.

More than half a lifetime watching birds at this single location has given him a unique insight into changes in the pattern of swallow movements. In 2019, Martin recalled – uniquely in all his time here – the first swallows arrived in mid-February, three or four weeks earlier than normal. This was, he explained, largely down to freakish weather conditions, with a broad swathe of warm air sweeping up towards Britain from north-west Africa, so these early migrating birds had just kept on going. Fortunately, we did not then experience a cold spell like the previous year's 'Beast from the East', so most of these early arrivals would have had a good chance of surviving.

Overall, Martin believes, swallow numbers are holding up well, especially when compared with the house martin. That species has hugely declined over the past three decades, with what he calls 'pitiful numbers' now passing through each spring.

He introduced me to Peter Morgan who in the early 1960s, as a keen but callow nineteen-year-old, was the very first warden of Portland Bird Observatory, before leaving to pursue a distinguished career as a university professor. Now, having retired, he is the chairman here. Like Martin, Peter thought swallows had managed to maintain their population over the past few decades, unlike many other migrants like the cuckoo and turtle dove. But he did suggest that more swallows are arriving earlier than they used to, and that there are now much bigger fluctuations in numbers. In some years, he told me, there are far fewer swallows passing through in the spring, possibly because of more severe storms in the Mediterranean.

I left Peter and Martin to their work and walked down the path, among what Martin calls the 'beach huts without a beach', to the sea. More swallows appeared: twos and threes, flying low over my head. At first, they were silhouetted against the brightening sky but, as they got closer, I glimpsed their pale underparts and blue-black upperparts as they twisted and turned in the wind. Some were so close I could see the shortened tail feathers that indicated they were youngsters, on their very first journey south. Would they make it? Their chances were less than fifty-fifty. Whether these birds were just feeding, or would soon head off, was hard to tell. Their flight always has a determined quality, especially on a gusty autumn day like this.

For the past six months, swallows had been my more-or-less constant companions. And just as when an old friend or family member is about to go away, so for a while I had been feeling an imminent sense of loss. It was just a few weeks before the clocks would go back. These windswept birds, I suspected, were the last I would see, for this year at least.

The next few days saw blustery storms come in from the Atlantic, bringing heavy and persistent rain, followed by, in late October, a cooler airstream from the north – the first breath of winter. Fieldfares chattered and redwings whistled in the clear blue skies above, while reports from my local birding website indicated the mass arrival of winter visitors – but not a mention of any lingering summer ones.

Occasionally, swallows are seen well into late autumn. One cold and murky afternoon in late November – the kind of day when I wished I'd put more layers on – I had come down to my coastal patch on the River Parrett with some birding companions for the falling tide, which would retreat across patches of mud ideal for waders. When the rain started, we turned for home. Then I saw a familiar silhouette: a lone juvenile swallow, rather bedraggled, its short tail showing it had only recently fledged and left the nest. Up and down it flew in the gathering gloom, as if unsure which way to go. There certainly weren't many insects to feed on. We were going back to our snug, warm homes; this little swallow had a long way to travel.

And there was one even later swallow. A decade or so ago, I received a phone message from our neighbour Mick, who tends his allotment across the road. Mick is a keen observer of birds, who once pointed me in the direction of an autumnal fall of whinchats and wheatears on the nearby moor, and is old enough to remember a time when 'cuckoos drove us mad' here in the village. Once I managed to get hold of him, he told me he'd been cycling past the field next door, when he had caught a brief but familiar movement of a small bird flying low over the grass.

I wandered over, not really expecting to see anything unusual. But there it was: perched on a drinking trough set into the hoof-trampled mud. A swallow, which then took to the wing and hawked around for any tiny insects still airborne. The date? 1 December.

That swallow was a real rarity: during fieldwork for the BTO's original *Atlas of Wintering Birds in Britain and Ireland*, published in 1986 but based on observations made from 1981 to 1984, fewer than 150 winter sightings of swallows were reported. Of these, the vast majority were in November or February, with just 13 records from the first few days of December which, like my bird, were clearly very late migrants. The most recent *Atlas* survey did produce a lot more winter records, mostly on or around our coasts, where temperatures tend to be higher than inland. 'Despite being a traditional harbinger of spring,' the authors concluded, 'swallows are departing later with growing frequency and in some areas attempting to overwinter.'

In January 2009, a lone swallow overwintering at Marazion Marsh in west Cornwall was given the nickname 'Rambo' for its extraordinary ability to survive temperatures of 12 degrees below zero. Yet the chances of these stay-at-home birds making it through to the following spring are – for the moment at least – vanishingly small.

Sometimes migrating swallows get caught out by unseasonable cold weather even before they have left. One of the pioneers of the study of animal behaviour, the Nobel Prize-winning scientist Konrad Lorenz, gave a detailed account of events in late

September 1931, in his home village of Altenberg on the Danube, in south-east Germany, following an unusually early cold snap:

> [Swallows] congregated in great masses … in the cracks and crevices of three houses. Hundreds and hundreds of swallows crowded into these apertures, the wings and tails of the latecomers protruding from the holes. Even outside the hole … a mass of birds lay on top of one another, their tails and pointed wings giving them the effect of a sea-urchin!

Lorenz naturally assumed that the swallows were dead, but, when he removed the outer layers of birds, some did respond and came to life, taking flight and darting off. They are not always so fortunate: during a long spell of very cold and wet weather in autumn 1974, many millions of swallows died in central Europe before they could even start to migrate.

Like other migratory birds, swallows find their way using a range of different cues. These include the Earth's magnetic field, using an internal 'compass' to orient them in the right direction: i.e. south in autumn and north in spring. But this only provides the general direction: like other migrants, it's likely they are aware of the position of the sun on clear days, and polarised light coming through the clouds on overcast ones. They will also follow landscape features such as mountains, rivers and coastlines and, as they near their destination, specific topographical features and even familiar smells.

In autumn, swallows move at a fairly leisurely pace: an average of roughly 15–16 miles (24–26 kilometres) a day, though this is achieved by travelling more on some days (up to 45 miles or 71 kilometres) and resting or feeding on others. But as autumn takes hold, and flying insects become scarcer, they quicken their pace, and can cover up to 120 miles (roughly 190 kilometres) a day. Surprisingly, perhaps, they prefer to fly into a headwind or crosswind rather than, as you might suppose, take advantage of a following wind; the same as we do when piloting a yacht. This is partly because it keeps them more stable, but also because it is easier to chase and catch insects when flying into a headwind.

Swallows migrate on a broad front, following different routes south depending on their starting point. Typically, birds from Britain and western Europe take a westerly route, heading across the Bay of Biscay, Iberia and the Mediterranean, while those from eastern Europe and western Asia take a different path, through the Middle East. It is often supposed that they avoid the Sahara Desert by heading along the coast, yet evidence shows that many do fly across at least part of it.

Nevertheless, the Sahara does present a major hazard: in *The Palaearctic-African Bird Migration Systems* the ornithologist Reg Moreau, who pioneered the study of these global journeys, noted that more barn swallows have been found dead there than any other species, and quotes a traveller 'who saw them fall dead in flight in front of him', presumably as a result of dehydration and overheating from the unforgiving sun.

Migrating birds will stop for shelter, food and water at desert oases. But if they can't, they will seek shelter anywhere: in the shadow of objects such as oil drums, among the ruins of former military buildings, even inside wrecked cars. One observer re-

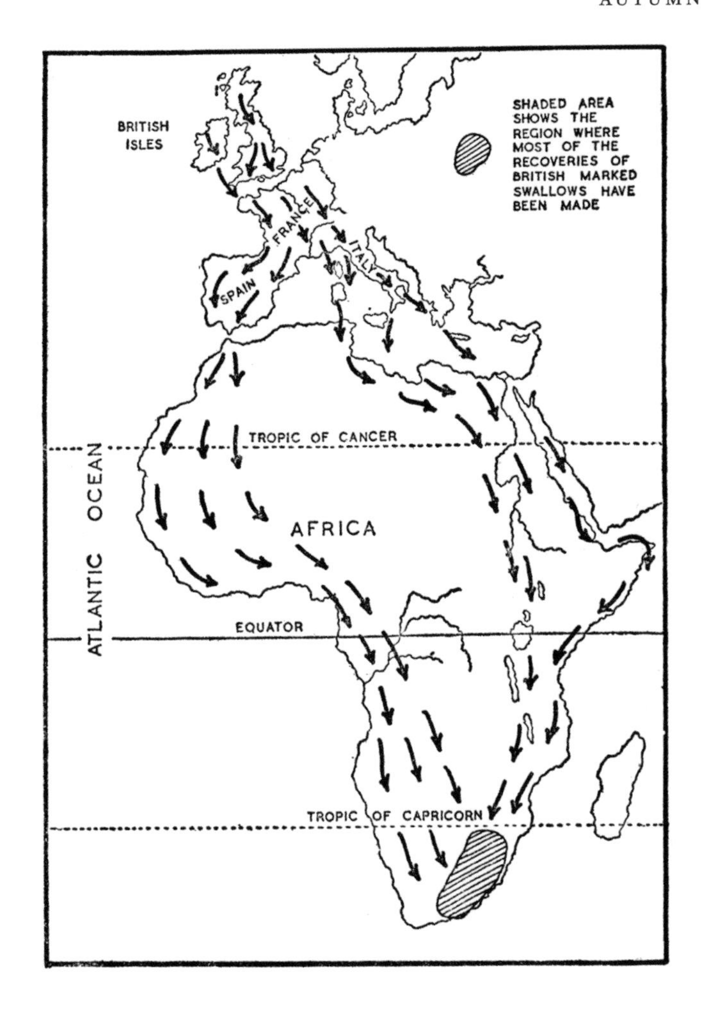

ported seeing two dozen turtle doves and a scops owl in a burnt-out vehicle, while several birds – including a swallow – sought sanctuary in his own car.

Unlike the majority of long-distance migrants, such as warblers, flycatchers and chats, swallows migrate not by night, but during the day, and at low rather than high altitudes, usually just a few metres above the ground. That is largely because they pause regularly along the way to feed. For this reason, swallows do not, unlike species such as sedge warblers, build up huge fat reserves before they depart south; however, they do put on some weight, so that if they encounter bad weather or a shortage of insects they still have enough energy to fly.

Because of their habit of migrating by day, they are observed more often than many other migrants, and are frequently seen in large flocks, sometimes numbering in the tens of thousands. These concentrations often occur when bad weather stops them in their tracks, or causes them to wait until it has passed. One such spectacle was observed on 1 November 1907 by the ornithologist and bird-collector Commander Hubert Lynes. Later in life Lynes rose to the rank of rear-admiral in the Royal Navy, serving with great distinction in both the First and Second World Wars, but in 1907 he was on the first of several expeditions to the area around Mombasa in what was then British East Africa (now Kenya). Lynes was understandably enthralled by the new and exotic species he was seeing: 'I was making acquaintance, to my great delight, with the Touracos, Hornbills, Sunbirds and other Ethiopian [i.e. sub-Saharan African] birds quite new to me, when about noon there flitted by two or three birds that looked very much like our English swallow.'

At first, he more or less ignored these familiar creatures: 'Beyond noting … the time and bearing in my pocket-book, I did not pay any particular attention, engrossed as I was with the African birds.' But as the afternoon went on, and more and more swallows began to appear, he turned his attention briefly towards them, shooting one specimen with what he called 'my little collecting gun' to make sure he had identified the species correctly. Then, the weather abruptly changed, with dark clouds betokening the approach of a rainstorm.

> The effect of this slight alteration of meteorological conditions upon the migration of the Swallows was wonderful. They obviously would not face the storm, and stopped. A banking up of the migratory stream resulted, fresh birds arriving continually from the north-north-east to swell the multitude, until the air was simply alive with Swallows wheeling around from the ground level to about 1,000 feet high, evidently catching flies and making the most of their time. There must have been thousands within my field of vision.

But then Lynes was distracted by what he called 'chasing some Irrisors' (green wood-hoopoes) and, when he looked again, a quarter-of-an hour later, almost all the swallows had gone.

> They had melted away, leaving hardly a single bird in view ... They had evidently continued their passage directly the threatening storm had cleared.
>
> My specimen proves to be an adult male in full plumage, with full chestnut forehead, cheeks and throat, dark pectoral band, and no sign of moult. It was very fat, and its stomach was full of flies, which it had no doubt caught just before I shot it.

Fortunately, swallows travelling south are no longer shot by gun-toting military men.

That autumn there was to be one last, fleeting encounter. In mid-October, a week after my farewell to the swallows at Portland, I went for a Sunday afternoon stroll at one of my favourite places, the RSPB reserve at Ham Wall near where I live. It had rained more or less the whole week and, what with the Rugby World Cup on the television, I realised I hadn't been out birding here for a while. But the morning rain had cleared, the sun had come out, and to my surprise the last dregs of summer had still not quite slipped away. On the path along the old railway, I came across red admirals and speckled woods sunning themselves on clumps of ivy, and a single painted lady – the first I had seen for several weeks – fluttering by.

Dragonflies were everywhere, too. Small and slender common darters – reddish males and greenish-yellow females – clustering in huge numbers on every available surface, or squabbling with one another over the path; and larger migrant hawkers, buzzing around to grab a snack from the clouds of midges hanging in mid-air, or clinging onto one another in a final mating embrace before winter set in.

Families were out, too, in the autumn sunshine, including three loud yet delightful children, with parents in tow. They apologised profusely as their youngsters bashed sticks and yelled with excitement; I told them how lovely it was to see children enjoying themselves outdoors – and I meant it, for they reminded me of my three teenagers when we used to come here in their younger days. Despite the warm weather, there were signs of winter too: whistling wigeon, newly arrived from Siberia, among the resident mallard and gadwall; two ravens cawing high overhead; and a mini-murmuration of starlings gathering far off. It really did feel as if this was summer's last hurrah.

And then I noticed them, high in the sky, flying determinedly towards me: a couple of tiny birds amidst the larger ducks and lapwings. Two swallows, surely the very last of the year, heading south into a light, warm breeze.

I was in two minds. My melancholy at their impending absence, which even the arrival of winter visitors like redwings and fieldfares couldn't quite diminish. But at the same time, I also felt – yet again – that wonder, which I rarely get with any other species, apart from perhaps the swift. These global wayfarers, forging their way into the autumn wind, would, that day, the next, or perhaps the following week, set out over the sea. And then over France and Spain; across the Mediterranean, the Sahara Desert, and equatorial Africa: to reach their winter home at the far end of this vast continent, more than 6,000 miles away.

After seeing my last swallow of the year, I don't usually encounter them again until one day, the following March or early April, I catch sight of a familiar shape flying low over my home village. But this year would be different. This winter, for the very first time in my half-century and more as a birder, I planned to follow them south.

GREETINGS from
PRETORIA
Police Station
Street
Valley near

5
WINTER

It has been proved that some British swallows pass the winter in South Africa, but we can tell few details of their journey thence and thither, or their escapes from wind and weather, from shotguns and birds of prey. We can only guess how they 'o'er Afric's sands careened' ... Yet our ignorance is not altogether regrettable; where knowledge is veiled, mystery still lives, and mystery is the mother of romance.

E. W. Hendy, *The Lure of Bird Watching* (1928)

It had been a long, tedious and exhausting flight. The night before, my plane had climbed out of a damp and chilly Heathrow, and now, at six o'clock on a bright summer's morning, I emerged from Durban's King Shaka Airport into heat, sunshine and a melee of indefinable yet tantalising sounds and smells. I had flown almost 6,000 miles and been travelling non-stop for almost 24 hours. I was very glad to be in Africa.

It was only as I was driving my rental car out of the airport that it dawned on me. I had been whisked here in a sealed metal capsule, with meals and movies served up at my convenience. Yet the swallow does this very same journey, from Britain to South Africa and back, year after year, under its own steam, propelled only by a diet of flies, for the whole of its life – all without the aid of modern technology.

Having followed my Somerset swallows to where they spend half their lives, I could now anticipate seeing them again. I also had to recalibrate my birding expectations: I had swapped buzzards for yellow-billed kites, dowdy house sparrows for bright, black-and-yellow weavers, and chunky starlings for elegant widowbirds, whose males were flying low across the adjacent fields trailing their impossibly long tails behind them.

The woman I was on my way to meet has done more than anyone to alert people to one of the world's greatest natural spectacles: a barn swallow roost. Angie Wilken emigrated to South Africa with her parents at the age of twelve. But she was born and raised in Brighouse in West Yorkshire, and still has something of the no-nonsense Yorkshirewoman about her. Thirty years ago, after getting married, Angie and her husband bought a building plot half an hour's drive out of Durban and moved out of the city in search of a quieter life. Mount Moreland felt surprisingly rural – 'We like to call ourselves the nearest faraway place!' she says.

Soon afterwards, while building their new home, they noticed that each spring and summer's evening the swallows were gathering to roost in huge numbers over a nearby reedbed, known as Lake Victoria. Since then she has adopted these birds as her own. 'It's something that reaches deep inside you,' she told me. 'They touch your emotion in a way you can't describe. And even though I see this spectacle every night, and have done for so many years, it's always somehow new and special.'

She was determined to share this discovery with others. 'I made enquiries with the local parks board, but nobody seemed to be interested. Then I had a vision: I rented the land and started inviting people to come and see them – and they loved it!' Angie built terracing on the slopes around the reedbed to create a natural auditorium, so that people could have a better view of the birds. At its peak, more than a thousand visitors a night were coming to see the swallow roost. As Burt Lancaster tells Kevin Costner in *Field of Dreams*: 'If you build it, they will come.'

But how many swallows? 'People estimate between three and five million,' Angie told me, 'but I prefer just to say it's uncountable.' Yet even these figures probably underestimate the total number of birds that benefit from this location: 'At the start and end of the season there is a real boost in numbers, which suggests that perhaps as many as ten million individual birds use the site at some point during the year.'

Ten million swallows equate to over 2 per cent of the *entire* global population of 190 million. No wonder that, in 2006, the site was designated as an Important Bird Area by the global conservation organisation BirdLife International. Angie should be proud of her lifelong project.

The swallows mostly arrive at Mount Moreland from mid-October, and the last birds usually depart around the Easter weekend. As to exactly where they start out, studies have shown that many of the birds that winter in KwaZulu-Natal – including Mount Moreland – come from eastern Europe and western and central Asia, with ringing recoveries from Finland, Germany, Hungary and Ukraine. Our birds, it seems, mostly head to the area around the Cape of Good Hope, or remain further north, in Namibia and Botswana.

However, the very first ringing recovery of a British swallow, in December 1912, did come from here, having originally been caught and ringed in Staffordshire by James Masefield, brother of the poet John. In the century or so since, a number of other British-ringed swallows have been recovered in this region, including two trapped by Andrew Pickles in the roost at Umzumbe, a couple of hours south of Mount Moreland. One, ringed in Powys, Wales on 24 June 2015 was recaptured at Umzumbe five months later, on 28 November; another, ringed on migration at Retford, Nottinghamshire, on 26 September 2011, travelled the 6,000-plus miles to South Africa in just two months.

'I wish they all had little flags,' Angie reflected, 'so we could tell exactly where each one comes from.' Soon, perhaps, her wish will be granted, as GPS systems and geolocators are now so light it would be no problem for a swallow to wear one all the way to Africa and back.

This was the largest known barn swallow roost in South Africa. Then in 2016, without warning, things began to deteriorate. The swallows only stayed for a few weeks, instead of the usual six months. The following year, they again arrived late and left early. In the 2018–19 season they stayed for just three days.

Angie was faced with a dilemma. If she kept the viewing site open, she ran the risk that visitors – some of whom might have travelled long distances – would be disappointed and angry. But by then the swallows had been part of her life for over twenty years, so she was understandably reluctant to call it a day. Eventually, though, she had no choice but to declare the site closed to visitors.

Yet to her – and my – astonishment, in October 2019, just three months before my visit, the swallows had returned in full force. Angie wonders if the recent heavy rains along the coast had made the roosts there temporarily unsuitable for the birds. David Allan, an ornithologist at the Durban Natural Science

Museum, agrees that rainfall elsewhere in the region might be a key factor. But he also points out that droughts further north in Africa could be pushing birds south, and that drier conditions around the Cape might also cause the swallows to concentrate in the east of the country.

In some ways, we should be grateful that the swallows are at Lake Victoria at all; for one major sporting event almost destroyed this roost forever.

On 15 May 2004 the President of FIFA, Sepp Blatter, revealed that South Africa would host the 2010 Soccer World Cup: the first time an African nation had ever been chosen. After the announcement, an emotional Nelson Mandela came onto the stage and raised the World Cup trophy aloft; something the South African team captain was unable to do six years later as, despite beating France in their final group match, they were knocked out at the first hurdle.

Hosting the World Cup was rightly seen as one of the key events in the rehabilitation of this troubled nation, after the dark days of apartheid. But with the city of Durban holding a number of key matches, including one of the semi-finals, the authorities decided it needed a new international airport. It would, they announced, be named after the early-nineteenth-century King Shaka – known as 'the black Napoleon' – who, depending on your view, was either the hero of the Zulu people or a genocidal tyrant responsible for wiping out millions of his enemies.

The site chosen for the airport – the same one I had just flown into – was less than two miles away from Mount Moreland, which meant that the flight path would pass directly over Lake Victoria. 'We fought a campaign against building the airport here at all, but sadly we lost,' Angie recalls. 'But we were successful in getting the authorities to install a "swallow radar", to ensure that there were no bird strikes.' Today, a decade later, the radar is still in operation, to safeguard both the birds and the passengers and crew of aircraft as they take off and land. Despite the ear-splitting noise as planes pass overhead, the swallows appear to tolerate their presence.

It could have been far worse. At one stage, the authorities were considering destroying the Lake Victoria reedbed to reduce the danger from bird strikes. However, a report demonstrated that, because swallows fly at much lower altitudes than any passing aircraft, the risk from them was minimal, so the site and the swallows won a last-minute reprieve.

Gathering in huge numbers at roosts like this is a clever survival strategy. For each individual bird, it radically reduces the chances of being caught by a predator. In the absence of suitable reedbeds, swallows roost in a wide range of habitats: from the branches of trees (mostly thick ones like acacia) to the ledges of buildings and telegraph wires in towns and cities – especially in the more populous areas of Asia. These urban roosts may not hold quite the vast numbers I was hoping to see here at Lake Victoria, but are nonetheless very impressive: one in Thailand's capital Bangkok has at times attracted almost 300,000 birds. Urban areas are doubly appealing to swallows: they are usually a few degrees warmer than the surrounding countryside, and the streetlights attract plenty of insects on which the birds can feed at dusk.

What struck me as soon as I arrived in South Africa was just how Anglocentric our attitudes towards swallows are. Ask anyone in the UK, 'Where do our swallows spend the winter', and the answer is likely to be, 'Africa'. Swallows do go to Africa, it's true; yet the question itself is misleading. As we've seen, they spend *our* winter in South Africa; yet in the Southern Hemisphere they are present during the austral spring and summer, where they are a ubiquitous presence from September through to April. Indeed, they are so familiar that the local nickname for rich Europeans who travel each year to their second homes around the Cape, to avoid the northern winter, is 'swallows'.

Barn swallows can be found across almost the whole country, in woodland, grass habitats and savannah – anywhere they can find flying insects – and all the way from sea level to an altitude of almost 3,000 metres (roughly 10,000 feet). They are, however, much more common in the wetter, eastern half of the country than in the drier west.

The numbers involved are enormous: it has been estimated that well over 100 million barn swallows enter sub-Saharan Africa each year, and they comprise the vast majority of all the hirundines there. One observer recorded all the swallows he encountered on a long drive across South Africa in February and March: he saw eight different species in total, of which 97 per cent were barn swallows. A reason for their ubiquity is that, unlike the species that breed here, which are tied to their nest sites, barn swallows can travel anywhere they like, in search of the most abundant supplies of food. The ornithologist Reg Moreau noted that in their winter quarters swallows are opportunist hunters: flying long distances to seek out places where ants or termites are emerging, or locusts are swarming.

Generally, swallows have a pretty easy life here, with plenty to eat. But in an extraordinary sequence filmed at Mapungubwe National Park, close to the border with Zimbabwe in the far north-east of the country, swallows feeding on insects above a river encountered an unexpected hazard. As they skimmed low over the surface, tiger fish leapt out of the water and grabbed the unsuspecting birds, dragging them under to their untimely death.

For the South Africans I talked to the arrival of barn swallows each October wasn't the same harbinger of their spring it is in the Northern Hemisphere – that honour goes to an intra-African migrant, the red-chested cuckoo – but they are intimately connected with the coming of rain. One couple, Mark and Lisa, told me that swallows often gather in large numbers just ahead of a rainstorm. But whereas back home in rain-drenched Britain, this might be viewed negatively, in South Africa it is seen as a positive sign: 'in Africa', Mark pointed out, 'rain is always welcome.'

Rain, or at least the period immediately after rain, means food. In southern Africa, studies have shown, barn swallows feed mainly on flies and beetles, though they also take a wide range of other large and small invertebrates, including spiders, grasshoppers, aphids, butterflies, moths and termites, which after a heavy rainstorm often swarm in their millions. Surprisingly, perhaps, they also occasionally feed on the fleshy outer part of some seeds – especially those of an invasive species of tree, originally from Australia, the coastal or cyclops wattle.

Swallows moult in Africa rather than Europe. This makes sense, given the abundance of food in this land of plenty, and that they do not have to be in peak condition outside the breeding season, as not having to feed hungry young means not needing to catch as many insects. But they cannot afford to replace all their feathers at once, or become temporarily flightless as some ducks do, as they must continue to fly and feed. So the whole process can take as long as six months.

During my brief stay in South Africa I saw barn swallows almost everywhere: perching on telegraph wires along the roadsides, swooping low around herds of game, or hawking acrobatically for insects above a nesting colony of Cape vultures at the top of the spectacular Oribi Gorge. Because they were moulting, they looked very different from the birds I see back home: rather tatty and tousled, with white specks on their blue upperparts. Most also had much shorter tails.

Despite the benevolent climate here in Africa, and the plentiful food, swallows do almost always depart once the austral autumn arrives, with the final birds usually leaving the roost by early April. However, Angie Wilken recalls one pair of barn swallows becoming stuck in netting at a sewage farm; having been rescued, they were kept in captivity to recover. When they were eventually released, they had missed the autumn departure of the other swallows, so remained here all winter. Other, free-flying birds have occasionally been sighted in the Western Cape in winter but, as far as we know, have never attempted to breed.

The evening after I met Angie, I arrived at Lake Victoria, a couple of hours before sunset, to witness the swallow spectacle for myself.

The view was not the prettiest I had ever seen: a skyline of distant hills topped with electricity pylons and the odd plume of smoke, in front of which was the dense reedbed where the birds would, I hoped, come to roost for the night. I was surprised how tiny it looked, for somewhere several million birds would soon be sleeping – the whole area is only about 10 hectares in all, about the same as 14 full-sized football pitches.

I was, I admit, rather nervous. The evening before I had visited a smaller roost a couple of hours to the south, near the Indian Ocean coast. Just before dusk, I had witnessed the arrival of several tens of thousands of swallows, but the occasion was far less spectacular than I had hoped. I had no guarantee that tonight would be any better, though I kept my fingers crossed that it would.

This evening, the sky was cloudy, and a light north-easterly breeze ruffled the tops of the reeds; a welcome relief to the stifling heat. The only sounds came from flocks of hadada ibis, reputedly the loudest bird in Africa, whose strident calls echoed in the distance; a different bird, which sounded remarkably like a song thrush, but clearly wasn't; and, every few minutes, yet another whining plane coming in to land.

Since Angie was still not publicising the birds' presence, I thought I would be on my own. Yet as time went by, I was joined by two or three groups of visitors, including a couple from Hampshire staying with South African relatives, who had heard about the roost and were curious to see if it was all it had been cracked up to be. They had even brought folding chairs and a picnic.

At half past five, I glimpsed two swallows as they passed briefly overhead. The first flocks didn't begin to arrive for another half an hour; I was relieved to see them, and once again thought of the long journey they had undergone to get here.

The light began to fade. A chorus of cicadas added their voices to the soundscape, while a strange, insect-like buzz emanating from the reeds turned out to be Pickersgill's reed frog, a minuscule amphibian less than three centimetres long, unique to the coastal lowlands of KwaZulu-Natal. How, I wondered, must more than a million swallows invading this tiny frog's home each evening trouble its consciousness? But as the philosopher Thomas Nagel pointed out in his essay 'What is it like to be a bat?', we cannot begin to understand another organism's experience – even when, as with the swallow, we somehow believe we can. They may inhabit the same world as us, yet they spend their lives in a parallel universe, which we are simply unable to enter.

My musings were interrupted by a brief movement at the back of the reedbed. Distances are deceptive here, and these birds were much further away than I had imagined; so much that, at first, I mistook them for insects.

But they were indeed swallows: several thousand of them, flying low over the fields of grass and sugar cane around the reedbed, still taking advantage of what remained of the light to feed. They were joined by a local species, lesser striped swallows, which flashed their orange rumps, and a handful of African palm swifts, whose elongated bodies made them appear even more emaciated than our own familiar species.

As we waited for the full spectacle, the picnicking party wandered over to ask me to take a group photo. I discovered, as I often did on this African odyssey, that as soon as I mentioned swallows, they straight away told me stories of their own. The lady mentioned that they owned a home in the South of France, where both swallows and house martins nested, and that she'd said goodbye to them a couple of months ago knowing that, like me, she would see them again far sooner than usual.

Three-quarters of an hour before sunset, a change came over the birds. They stopped flying low over the fields, and began to ascend into the air, starting to behave less as individuals and more like loose groups. It felt as if the main event was about to kick off.

Five minutes later, as I panned across the skyline, I could see hundreds of thousands of tiny specks, like midges. Closer in, flocks were gathering over the reedbed itself, flying around aimlessly as if killing time before taking the decision to land. As with any roosting behaviour, this is a complex game of bluff and counterbluff: the first bird that decides to drop down into the reeds may be immediately followed by thousands more, guaranteeing that it gets the best space at the centre of the mass. On the other hand, it may find it has gone too soon, and is alone and vulnerable to any lurking predator, while its companions remain airborne.

Minute by minute, the numbers continued to build, as hundreds of thousands more birds arrived. Looking at a bright circle of sky through my telescope, my retina could hardly cope with the complexity of patterns, as myriad specks swept up and down, swirled and poured in on themselves, bulged and sagged, in three very crowded dimensions.

I lifted my eyes to take in the whole scene. Clouds of swallows were swarming low above my head, increasingly anxious to reach the roost before nightfall. The other species had disappeared: every bird I could see was a barn swallow, as if confirming this one species' global dominance over its cousins.

And then I heard, for the first time for months, a familiar twittering sound: the sound of a summer's day in my Somerset garden; that magical 'insect music'.

I found it hard not to compare what I was seeing with another natural event, the winter roost of starlings on the Avalon Marshes in Somerset. There, you can stand and watch without optical aids: indeed, it is often better to, because you can take in the whole aerial pageant, as the starlings twist and turn in the evening sky. Here, the swallows stretched across such a huge area that with the naked eye it was hard to see them at all: only when I scanned with telescope or binoculars did I appreciate the magnitude of what I was witnessing. It was as if the entire firmament was alive: millions of tiny organisms moving to and fro; each one an individual swallow but collectively, as Alexander Skutch wrote in *The Minds of Birds*, 'like a huge aerial amoeba, with internal particles in constant agitation'.

With thirty minutes to go to sunset, a phalanx of swallows plunged down towards the reeds. At first, I assumed they were landing, yet again they did not: this was just a final recce.

By now the sky looked as if it had been invaded by tiny tornadoes: kettles of birds reaching up from the earth to the heavens, everywhere from a few centimetres above the reeds to so high they were almost out of sight. A few did then briefly land, trembling momentarily on a reed stem before taking to the wing again as if they had received an electric shock. Their tentativeness seemed uncharacteristic of such a usually confident bird. As always when watching any natural spectacle, I was torn between awe and curiosity. How do these birds know where to go? How do they learn exactly when to come down to land? And why were some now safely ensconced in the reeds, while others were still flying about? Where, in short, does the will of the individual swallow become that of the whole vast flock?

As the sky began to darken, tens of thousands of swallows dropped rapidly into the reeds, some closing their wings and plummeting down as fast as possible, others descending more slowly to settle on bended stems.

I had seen what I had come to see.

But I had one further ambition before I left South Africa: to see another species of swallow; one that, in contrast to its cousin, is so rare that even many local birders have never set eyes on it. On my quest, I was accompanied by the Lee family: Craig and Chantelle, and their children Wade and Casey – all keen birders, who I had met through a mutual friend.

This was our second visit to Roselands Farm, a landscape of

rolling hills, grasslands, pools and streams near the town of Hella-Hella, about an hour-and-a-half west of Durban. These disconnected patches of habitat, found only in this particular part of KwaZulu-Natal, and roughly 1,000 metres (3,280 feet) above sea level, are known as the 'mistbelt grasslands'. Indeed, two days earlier, after a long and tricky drive through mist and heavy rain, we had been greeted with such terrible weather that no self-respecting swallow of any species would take to the air, so we'd headed straight back home.

But today it was bright and warm: ideal conditions to search for our target bird. A fine-looking, fox-coloured Ridgeback dog, a large and formidable breed traditionally used to scare off lions, bounded up to say hello, accompanied by the farm manager, Zelt. We asked Zelt whether many people came to see these special birds.

A few years ago, he told us, four German guys had turned up in a hired Mercedes fresh off the overnight flight, and were duly pointed in the right direction. An hour or so later they came back. Zelt offered them a coffee and asked where they were headed next. They politely declined: now they'd seen the birds, they explained, they had to get their flight back to Germany. Serious dedication to global twitching – or complete insanity.

We may not have been quite so single-minded, but we really did want to catch up with these rare swallows. Craig was especially tense, having lived just a few hours away for many years, but never seen them. We both knew that, however 'guaranteed' a bird might be, there is always the chance you won't find it. The five of us piled into the back of a Toyota Hilux pick-up, or 'bakkie', as they are known here, with our guide, Sbu, who drove us on a short but bumpy ride to a grassy valley on the edge of the farm. We hopped off the truck and scanned.

At first, nothing. Then – a moment of panic and excitement – a pair of all-dark swallows appeared over the horizon and swooped down into the valley. But a closer look revealed them to be black saw-wings, a common species we had already encountered several times.

Then seventeen-year-old Wade's sharp eyes spotted another, more graceful movement. 'There it is! There it is!' he shouted.

To our joy and delight, there it was. Skimming along, barely a few centimetres above the waving grass, was one of the most beautiful birds I have ever seen: rich, dark, iridescent

purplish-blue, glinting in the sun, and trailing an impossibly long tail. It all added up to the sleek shape – at once familiar, yet strangely unfamiliar – of a blue swallow.

This was no ordinary hirundine, but one of the rarest and most sought-after birds in the whole of southern Africa. Birders' fascination with this species is partly down to rarity, but also because of its unique appearance: entirely steel-blue, with a distinctive metallic sheen when it turns towards the light.

Blue swallows are also incredibly buoyant in flight: weighing just 13 grams (less than half an ounce), yet longer than most other members of their family, thanks to their extended tail streamers, which on the male are noticeably longer than the body. They don't fly, exactly, but simply float just above the ground, as if suspended by magic.

Also known as the montane blue swallow, this upland species can occasionally be seen on migration across a surprisingly wide area of this vast continent, from here in KwaZulu-Natal northwards through Mozambique, Zimbabwe, Zambia, Malawi and Tanzania, to the Democratic Republic of Congo. All – including the South African birds – head north outside the breeding season, to concentrate in one small area around Lake Victoria (the huge waterbody on the Equator, not the reedbed where I saw the barn swallow roost).

Yet although the blue swallow's breeding range covers more than half a million square miles (1.3 million square kilometres) – more than five times the size of the UK – these birds are spread so thinly that they are very hard to track down. With a tiny global population of fewer than 2,500 individuals, and possibly even fewer, it is no wonder that BirdLife International has classified the blue swallow as Vulnerable: just three steps away from extinction.

To put this into perspective, the entire global population of blue swallows is roughly a thousand times smaller than the number of barn swallows I had seen a few days earlier at Mount Moreland, and a minuscule fraction of that species' world population. The cause of the blue swallow's decline is the usual story: the ongoing loss and fragmentation of its specialised habitat, so that only a few fragments of mistbelt grassland – less than 1 per cent of the original area – now remain. Another possible reason is the fickle weather here – even, as we had discovered, at the height of summer – making the supply of flying insects often erratic. If mist blankets these hillsides for even a day or two when the chicks are in the nest, they are likely to starve to death.

Yet these birds still bring delight to anyone fortunate enough to see them. Sbu (short for Sibusiso, which means 'blessing') has been showing blue swallows to visitors here for seven years now, and never tires of it. 'When they arrive back in spring, and I see them flying towards me and flashing those dark blue wings, I feel such joy.'

Blue swallows are unusual in that they are thought to be co-operative breeders: more than one male and female attending a single nest. 'As with most types of co-operative breeding', notes ornithologist David Allan, 'the "extra" individuals most likely represent offspring from earlier breeding attempts by the breeding pair, rather than several "pairs".' They make their nests either in aardvark burrows or sinkholes, building a half-cup of mud, grass and other vegetation against the side wall, and lining it with white feathers so they can find the nest in the semi-darkness. Later we watched as the female of this breeding pair (identifiable by her noticeably shorter tail) grabbed a feather floating in mid-air, dropped it, and then turned back on herself to catch it again before it hit the ground.

We spent a memorable morning observing the blue swallows come and go, and occasionally rest momentarily on a dead stick before taking to the air again. I love all swallows, but I cannot recall watching such effortless ease as this bird cruised the air, its all-blue plumage gleaming in the strengthening sun.

When it was time to leave, we stopped on top of the ridge and looked back. In the distance a single blue swallow was still feeding low over the grass. I had the sinking feeling I would never see this beautiful bird again.

The more I think about it, the more I am struck by the contrast between these two closely related species. Both barn and blue swallows have adapted their body shape and behaviour to feed on flying insects; both have found places to nest, in either natural or man-made structures; both have entranced generations of human observers. But the barn swallow is globally widespread and highly successful, while the blue swallow is rapidly heading towards oblivion.

But might I be focusing too closely on the plight of the rare species, and failing to appreciate the problems faced by the common one?

The barn swallow is, as we have seen, a global voyager: one of a tiny handful of species to have been seen on all seven continents, and familiar to people in most countries around the world. Its ability to constantly seek out new horizons, like the bird I encountered in Iceland, or the one that made it all the way to Antarctica, is one crucial secret of its success. The other has been its ability – along with only a handful of other bird species – to live so closely alongside human beings.

Yet as recent events at the Mount Moreland roost have shown, we cannot afford to be complacent about the fate of the barn swallow, even though, for now at least, it remains one of the commonest, most widespread species of bird in the world. Abundance is no guarantee of survival, as we learned with the passenger pigeon and Eskimo curlew: as recently as the second half of the nineteenth century, both species were so common as to be virtually uncountable; yet hunting and habitat loss saw both rapidly become extinct.

It's hard to imagine anyone hunting a bird as small and light as a swallow for food; and yet, when they form these vast night-time gatherings, they do become an easy target. In Nigeria, barn swallows have been captured at their roosts for decades, using a sticky substance smeared over the vegetation to trap the birds as they come down to land. Roosts are sometimes deliberately burned, or the crops where the birds land are harvested, so they have to move elsewhere.

What else might come to threaten barn swallows? A major problem is the recent catastrophic decline of insects and other invertebrates – memorably dubbed by conservation scientists 'Insect Armageddon'. When, in autumn 2017, the headlines announced that 75 per cent of flying insects had disappeared in just a quarter of a century, for once the media could not be accused of exaggerating. For what was really shocking is that this decline had occurred not throughout a mixture of good and not-so-good habitats, but on nature reserves across Germany, where the study had taken place. If insect numbers have dropped by three-quarters on sites specifically protected for wildlife, how badly must they be faring in the wider countryside?

There are many reasons behind this unprecedented and frankly terrifying decline: the over-use of pesticides, summer droughts, and habitat loss in the areas surrounding the reserves, have all been implicated. But for the barn swallow, a future problem is likely to be wholesale changes in cattle farming. The image of the traditional farmyard, surrounded by fields where contented cows graze on grass to produce milk or meat, is rapidly becoming out-of-date.

Nowadays, many cattle are being raised indoors, on what are known as 'mega-farms' housing up to 3,000 animals on a single site – in what amounts to factory-farm conditions. Indoor cattle may attract some insects, but these are unlikely to be accessible to feeding birds; nor do these places have any room for nesting swallows.

The threat to swallows from modern farming methods goes back a long way. As early as 1942, in his multi-volume work *Life Histories of North American Birds*, the ornithologist Arthur Cleveland Bent celebrated the close connection between swallows and farmers:

> No bird in North America is better known as a welcome companion and a useful friend to the farmer, as it comes each spring to fly in and out of the wide-open barn door, delighting him with its cheerful twittering, or courses about the barnyard in pursuit of the troublesome insects that annoy both man and beast.

However, he sounded a prescient warning about the future of this age-old relationship:

> But such a charming rural scene is not so common as it used to be. The old-fashioned barn, with its wide-open doors ... and the open sheds where the farm wagons stood are being replaced by modern structures ... with tightly closed doors and no open windows through which the birds can enter. Horses are replaced to a large extent by automobiles and tractors; cattle are housed in modern dairy barns ... There is no room for the swallow in modern farming.

The greatest existential threat to all living things, including, of course, humanity, is the global climate crisis. Reducing carbon in the atmosphere can be done in a number of ways, but one of these is to put much stricter limits on the number of cattle, which, via their digestive system, produce the damaging 'greenhouse gas' methane. The incipient shift away from consuming meat and dairy products, towards a more vegetarian (and increasingly, vegan) diet, will undoubtedly help combat climate change, but could spell trouble for the barn swallow, whose ability to exploit our traditional farming methods has, until now,

been the key to its global success. Ironically, a move away from livestock farming could cause the swallow to spiral into a rapid, even potentially irreversible, decline.

Meanwhile, though climate change is likely to allow swallows to extend their breeding range northwards, it will make some of their current breeding areas unusable; whether the benefits will outweigh the drawbacks is still up for debate. What is, however, certain is that the increasing desertification of many areas on the swallows' migration routes – including the likely expansion of the Sahara Desert – can only be bad news.

Climate change is also significantly altering the timing of the arrival and departure of swallows in the UK. According to long-term studies by the British Trust for Ornithology (largely fuelled by data collected by amateur birders), swallows are now arriving each spring a full three weeks earlier than they did in the 1970s. In 2019, many returned even earlier, with widespread reports of sightings in the last week of February. 'Swallows return early,' ran the headline in the *Daily Telegraph*, 'as Britain basks in summer-like temperatures.'

In the short term, an earlier arrival could temporarily increase swallows' breeding success, as it might enable them to raise three broods in a single summer. But in the longer term it is likely to spell disaster. The nightmare scenario is that our spring weather gets even more unpredictable, with unseasonably early heatwaves followed by cold snaps, killing the insects on which these early-returning swallows feed.

In the longer term, should the climate emergency not only wreak havoc on nature, but on humanity too, where would we be? Perhaps we should heed a warning from the nature writer and conservationist W. H. Hudson. More than a century ago, in 1895, he postulated a dystopian future with no room for either swallows, or us: 'There would be few swallows in a dispeopled and savage England, with all its buildings crumbled to earth.'

Almost a century later, in the 1991 book *Romantic Ecology: Wordsworth and the Environmental Tradition*, the biographer and critic Jonathan Bate had similar fears about a future without swallows, if climate change does its worst:

> Keats's ode 'To Autumn' is predicated upon the certainty of the following spring's return; the poem will look very different if there is soon an autumn when 'gathering swallows twitter in the skies' for the last time.

*

Just as I'd been about to leave the swallow roost back at Lake Victoria, something rather wonderful happened.

The sun was setting below the horizon, its rays shining through beneath the cloud base to illuminate the fringes of each cloud, as though they had been edged with gold leaf. Against this radiance, the swallows were silhouetted, twisting and turning in the dying light in their farewell dance. All around me was awash with swallows. One million? Two million? Three? I could not begin to tell.

Then, with a sense of urgency they had not shown till now, the latecomers folded their wings and plunged like guided missiles down into the reeds.

The western sky was still glowing orange well after the sun had finally set, but all the birds were down. Had I not witnessed their descent for myself, the only clue to their presence would be a soft murmur, like a rushing wind, from the depths of the reedbed. It was the sound of millions of swallows preparing to go to sleep.

EPILOGUE

Fly away, fly away over the sea,
Sun-loving swallow, for summer is done;
Come again, come again, come back to me,
Bringing the summer, and bringing the sun.

Christina Rossetti,
'Fly Away, Fly Away' (1862)

The swallow knows no borders or boundaries. It crosses countries and continents, bringing joy to the lives of so many people. Whether we live in the Northern or Southern Hemisphere, it reconnects us, each spring, with the natural world. Yet right now, millions of people prefer to fuel their insular, nationalistic prejudices with hostility towards the foreign and unfamiliar. Why, I wonder, are we so obviously unable to learn any lessons from how this globetrotting bird lives its life?

As our world becomes ever more urbanised, and billions of new city-dwellers become disconnected from the rhythms of the natural world, the swallow is also beginning to lose the significance it once had. 'We now have a different relationship with swallows from that of previous generations,' observes Angela Turner:

> Although many people still welcome them as a sign of spring, most of us, at least in Western society, are no longer bound by religious or superstitious needs to protect them. When they fail to return to our barns, houses and gardens, we may no longer fear that it is a sign of bad luck. But it is perhaps a sign that all is not well with our own environment.

The swallow is still, as always, a messenger: but now its message is not of joy, but of doom. It is the 'canary in a coal mine' of our time: warning humanity of the potential harmfulness of, for example, radically altering the world's climate. In future summers, to paraphrase Aristotle, one swallow may be all we can wish for.

And yet I cannot help but be optimistic. The very existence of the swallow – a bird that symbolises so many of our most important values, including freedom, resilience and, above all, hope – must surely be a sign that, even at this eleventh hour, all may not yet be lost.

So once again, I am waiting. And once again, I am not alone. The swallow may not be on everyone's seasonal radar, but I know I share my nervous anticipation with millions of others. On the day of the spring equinox, the annual moment when, here in the Northern Hemisphere, the Earth tilts us from winter into spring, we are all waiting. Maybe the first swallow will return today, or tomorrow, or, if the weather is bad, next week. But we do know that for now, at least, it will return.

STEPHEN MOSS

Mark, Somerset, March 2020

Acknowledgements

Huge thanks to everyone at Square Peg (Penguin Random House) involved in this book, who continue to meet such high standards: Mireille Harper in Editorial; Ryan Bowes in Publicity; Jane Kirby in the rights team; and the designers and production team, Rosie Palmer, Lily Richards and Shabana Cho, who found the lovely cover image by Malcolm Greensmith. Thanks also to Sarah-Jane Forder for such thorough proofreading. Special thanks go to my wonderful editor, Rowan Yapp, who is now moving on to new and exciting challenges. And my gratitude, as always, to my dear friends Graham Coster, who edited the manuscript, and Broo Doherty, my agent.

Martin Cade and Peter Morgan of Portland Bird Observatory provided fascinating information on swallow migration and behaviour, as did Dawn Balmer at the BTO, Jack Baddams at the BBC Natural History Unit, Derek Niemann, Andrew and Duncan Ratcliffe of Ratcliffe Motors, and Andrew Graham-Brown of AGB Films, who allowed me to watch his wonderful footage of nesting swallows. My colleague at Bath Spa University, Gail Simmons, kindly read through the text.

Unlike my previous books on the robin and the wren, *The Swallow* involved a trip abroad, to South Africa, where our swallows go for half the year. This would not have been possible without generous help and assistance from the following people: South African Tourism (Rachel Lewis, Rebecca Riley and Grace Armitage), the Nicholson family and their staff, Keith Kopman and Crystal Brook at Rockjumper Bird Tours, Andy Ruffle, Andrew Pickles, and especially Angie Wilken, whose efforts at Mount Moreland over so many years have helped safeguard the huge roost, and allow so many people to witness this incredible spectacle. On my trip to see the blue swallows I was joined by Craig, Chantelle, Wade and Casey Lee. David Allan of the Natural Science Museum, Durban, kindly read through my drafts and made many helpful comments.

Adam Riley of Rockjumper hosted me with great kindness and generosity; my special thanks to him, his wife Felicity, their delightful children Will, Alex and Tori, and all their friends, for making my visit to South Africa so special.

Most of all, I owe a huge debt of thanks to Angela Turner, who knows more about swallows than almost anyone, and generously shared her expertise, knowledge and enthusiasm with me.

List of Illustrations

H. Grønvold pinxt.

Bemrose & Brailsford Lith. Sheffield.

PASSERIFORMES: HIRUNDINIDÆ, MUSCICAPIDÆ.